Lecture Notes in Mathematics 1482

Jan Chabrowski

The Dirichlet Problem with L^2-Boundary Data for Elliptic Linear Equations

Springer-Verlag
Berlin Heidelberg New York
London Paris Tokyo
Hong Kong Barcelona
Budapest

Author

Jan Chabrowski
Department of Mathematics
The University of Queensland
St. Lucia QLD 4072, Australia

Mathematics Subject Classification (1991): 35B, 35D, 35J

ISBN 3-540-54486-0 Springer-Verlag Berlin Heidelberg New York
ISBN 0-387-54486-0 Springer-Verlag New York Berlin Heidelberg

Printed in Germany

Typesetting: Camera ready by author
Printing and binding: Druckhaus Beltz, Hemsbach/Bergstr.
2146/3140-543210 - Printed on acid-free paper

Contents

INTRODUCTION

The purpose of this work is to give a self–contained study of the Dirichlet problem with L^2–boundary data for linear elliptic equations. In the last decade this problem has attracted the attention of several mathematicians, giving rise to interesting ideas and methods. We are concerned with the solvability of the Dirichlet problem

$$-D_i(a^{ij}(x)D_ju + b^i(x)u) + d^i(x)D_iu(x) + c(x)u = f(x) \text{ in } Q,$$

$$u(x) = \varphi(x) \text{ on } \partial Q,$$

where Q is a bounded domain in R_n.

In the case where φ is the trace of an element of $W^{1,2}(Q)$ this problem has an extensive literature (see **[LM]**, **[LU]**, **[GT]**, **[ST1]**, **[MI]**, **[M5]**). In this work we assume that $\varphi \in L^2(\partial Q)$. It is known that not every function in $L^2(\partial Q)$ is a trace of a function from $W^{1,2}(Q)$. Consequently we cannot expect to find a solution in the space $W^{1,2}(Q)$. It turns out that a suitable Sobolev space is $\widetilde{W}^{1,2}(Q) = \{u;\, u \in W^{1,2}_{\text{loc}}(Q),\, \int_Q |Du(x)|^2 r(x)\, dx + \int_Q u(x)^2\, dx < \infty\}$, where $r(x) = \operatorname{dist}(x, \partial Q)$. Also, the boundary condition requires a new interpretation. To give a new meaning to the boundary condition, we consider traces of a solution on surfaces parallel to the boundary ∂Q. Under some regularity assumption on ∂Q, these traces have limit in the sense of L^2–convergence, as these surfaces converge to the boundary. Adopting this approach, one can justify the use of the weighted Sobolev space $\widetilde{W}^{1,2}(Q)$ to solve our problem. On the other hand this approach allows us to derive a necessary and sufficient condition for a solution to have a trace on the boundary. We point out here that the Dirichlet problem with L^2–boundary data originates in the theory of analytic functions. We say that the analytic function $f(z)$ defined on $\{|z| < 1\}$ has a limit in L^2 on the boundary, if there exists a function $\varphi \in L^2(2\pi)$ such that

$$\lim_{r \to 1-0} \int_0^{2\pi} |f(r \exp i\theta) - \varphi(\theta)|^2\, d\theta = 0.$$

F. Riesz (see **[RE]**) proved the following criterion: the analytic function $f(z)$ on $(|z| < 1)$ has a limit on the boundary if and only if the function

$$U(r) = \int_0^{2\pi} |f(r \exp i\theta)|^2\, d\theta,\ 0 \leqq r < 1,$$

is bounded. Later, Littlewood and Paley (see **[LP]**) established the following criterion: the analytic function $f(z)$ on $(|z| < 1)$ has a limit in L^2 on the boundary if and only if the function

$$g(\theta) = \left[\int_0^1 (1 - r)|f'(r \exp i\theta)|^2\, dr\right]^{\frac{1}{2}},\ 0 < \theta < 2\pi$$

belongs to $L^2(0, 2\pi)$. The results of Riesz and Littlewood–Paley were extended by many authors to harmonic functions (for historical material and bibliographical references see the monograph [KR]). It appears that first extensions of these results to elliptic equations can be found in papers by Nečas and Mikhailov (see [MG], [M1–4], [N1–2]). In these papers this problem was solved under the assumption that $a^{ij} \in C^1(\bar{Q})$. This assumption is needed to recover the boundary data via Green's theorem. Hoffmann-Walbeck overcame this difficulty by a suitable mollification argument and was able to show this for coefficients a^{ij} satisfying the Dini condition (see [HW1–2]).

The plan of this work is as follows. Chapter 1 provides all relevant information from the theory of elliptic equations. In particular we discuss the solvability of the Dirichlet problem in $W^{1,2}(Q)$. We stress here that to solve the Dirichlet problem with L^2–boundary data, we need the weighted Sobolev space $\widetilde{W}^{1,2}(Q)$, which obviously contains $W^{1,2}(Q)$. We reproduce an argument, due to Poulsen (see [PO]), showing the existence of functions in $\widetilde{W}^{1,2}(Q)$, which do not have traces on the boundary.

In Chapter 2 we solve the Dirichlet problem with L^2–boundary data in a half space, under very weak assumptions on the coefficients a^{ij}. A Dini – type condition is only needed for coefficients a^{ni} in the normal and tangential directions to the boundary. Theorem 2.3 gives a necessary and sufficient condition for a solution in $\widetilde{W}^{1,2}(R_n^+)$ to have a trace on the boundary. This condition can be expressed by one of the following conditions

(I) $\int_{R_{n-1}} u(x', \delta)^2\, dx' < \infty$ is bounded for $\delta \in (0, \delta_\circ]$

and

(II) $\int_{R_n^+} |Du(x)|^2 \min(1, x_n)\, dx < \infty$.

In fact both conditions are equivalent. These conditions play a significant role in our approach to the Dirichlet problem. Any solution of an elliptic equation satisfying one of these conditions has a trace on $x_n = 0$.

In Chapter 3 we solve the Dirichlet problem with L^2–boundary data for a bounded domain Q with the boundary ∂Q of class C^2. The mollification argument from Chapter 2 is carried out by flattening locally the boundary and then by using a partition of unity; we define coefficients A^{ij} which are of class C^1 on Q and approximate a^{ij} in some sense. The conditions (I) and (II) for bounded domains take the form

$$\sup_{\{0<\delta\leqq\delta_\circ\}} \int_{\partial Q_\delta} u(x)^2\, dS_x < \infty \text{ and } \int_Q |Du(x)|^2 r(x)\, dx < \infty,$$

respectively, where $Q_\delta = \{x \in Q;\ \mathrm{dist}(x, \partial Q) > \delta\}$. Since ∂Q is of class C^2, there exists a one-to-one mapping $x_\delta : \partial Q_\delta \to \partial Q$. We show that any solution u satisfying one of the conditions (I) or (II) has a trace on ∂Q in the sense that x_δ converges to some $\varphi \in L^2(\partial Q)$, as $\delta \to 0$. The conditions (I) and (II) lead to the definition of a weighted Sobolev space in which we establish the existence of a solution to the Dirichlet problem with L^2–boundary data. Since the space $\widetilde{W}^{1,2}(Q)$ contains functions which do not have traces on ∂Q, Theorems 2.3 and 3.4 show that the property " function $u \in \widetilde{W}^{1,2}$ has a trace on the boundary " is a property of a solution not of the space $\widetilde{W}^{1,2}$. The proofs of the existence of a solution to the Dirichlet problem in Chapters 2 and 3 are based on the energy estimate. A significant feature of the energy estimate is that it involves the norm of a boundary data in $L^2(\partial Q)$.

Chapter 4 is devoted to the estimates of derivatives of a solution of the Dirichlet problem with L^2–boundary data. In particular we show that

$$\int_Q |D^2u(x)|^2 r(x)^3 \, dx < \infty$$

and we also derive a higher integrability property of Du. We use some ideas due to Iwaniec (see [**IW**]).

The results of Chapter 3 allow us to show in Chapter 5, that harmonic measure relative to elliptic operator is mutually absolutely continuous with respect to the Lebesgue surface measure. This result is established under the assumption of a Dini condition on a^{ij}. This assumption is very close to the optimal one since there exists an example due to Modica and Mortola (see [**MM**]), of an elliptic operator with $a^{ij} \in C(\bar{Q})$, for which the harmonic measure is singular with respect to the Lebesgue measure. We shall return to this question at the end of Chapter 10, where we show that in fact the Dini condition for a^{ij} is needed only on ∂Q.

The results of Chapter 3 make it possible to study the question of the uniqueness of the Dirichlet problem. In Chapter 6 we show that a solution assumes boundary values on ∂Q except for a set $E \subset \partial Q$. We prove that the space L^p, with $\frac{n}{n-1} \leqq p < \infty$, is a class of uniqueness for the Dirichlet problem if E has a finite Hausdorff measure of order $n-q$, where $\frac{1}{p} + \frac{1}{q} = 1$. The results of this chapter are due to Gajdenko (see [**GA**]).

In Chapter 7 we show how the methods developed in Chapter 3 can be used to solve the Dirichlet problem for a degenerate elliptic equation. The coefficients a^{ij} of the equation in question vanish, identically on ∂Q. The lack of ellipticity on ∂Q is compensated by a rather strong assumption on the coefficients of $D_i u$. In the final part of this chapter we apply our L^2–approach to solve some non–local problem in the sense of Bitsadze–Samarskii.

The main objective of Chapter 8 is to show the use of a regularized distance in our L^2–approach to the Dirichlet problem. This allows us to weaken the assumptions on the boundary ∂Q, namely we assume that ∂Q is of class $C^{1,\alpha}$. Under this assumption, the mapping x_δ can only be defined locally. On the other hand $r(x)$ may not be of class C^2 near the boundary. This was the key property used in our approach to the Dirichlet problem. To overcome this difficulty we use the concept of regularized distance ρ. The second derivatives of ρ behave near ∂Q like $\rho^{-\beta}$ for some $0 < \beta < 1$. This is sufficient to extend our method to $C^{1,\alpha}$–boundaries and even to $C^{1,\mathrm{Dini}}$. The results of this chapter are patterned after works of Petrushko and Lieberman (see [**PE**] and [**LI1-2**].

As we pointed out in Chapter 1, functions belonging to the space $\widetilde{W}^{1,2}(Q)$, in general, do not have traces. In Chapter 9 we present the recent results of Guščin [**GU**] who constructed a space $C_{n-1}(\bar{Q})$ containing $W^{1,2}(Q)$, such that every function from this space has a trace in $L^2(\partial Q)$ and every function in $L^2(\partial Q)$ can be extended to a $C_{n-1}(\bar{Q})$–function on Q. In this chapter we present a construction of the space $C_{n-1}(\bar{Q})$ based on some properties of Carleson measures. We also prove some characteristic properties of C_{n-1}–functions. These properties show that $C_{n-1}(\bar{Q})$ is close to the space $C(\bar{Q})$. Elements of the space $C_{n-1}(\bar{Q})$ have a continuity property which is analogous to the continuity of functions of several variables. This is one of the reasons that Guščin calls the elements of $C_{n-1}(\bar{Q})$ "the $(n-1)$–dimensionally continuous functions".

The final Chapter 10 contains another approach to the Dirichlet problem on domains with $C^{1,\mathrm{Dini}}$–boundary. We present here the approach due to Guščin [**GU**]. This question has already been discussed in Chapter 8; however we reproduce here Guščin's approach since it is vital in proving C_{n-1}–estimate of the solution of the Dirichlet problem. We consider the operator L containing only the leading coefficients a^{ij}. It is assumed that a^{ij} are in $L^\infty(Q)$ and satisfy the Dini condition only on the boundary. This is a significant relaxation of the regularity assumption on a^{ij} compared with Chapters 3 and 8. We establish the energy estimate by constructing the coefficients $a_\circ^{ij}$ which are close to a^{ij}, allowing us to use the Green theorem to recover boundary data. However, this can be done locally. The construction of approximating coefficients A^{ij} in Chapter 3 is more involved but we can define them globally. Also, due to this fact it is easy to handle equations containing lower order coefficients. This seems to be more difficult under the assumptions on a^{ij} in Chapter 10. Certainly, the energy estimate of Chapter 10 continues to hold for operator L with bounded coefficients b^i and c. Finally, we discuss the most striking feature of the space $C_{n-1}(\bar{Q})$. Namely, the C_{n-1}-norm of a solution of the Dirichlet problem with L^2–boundary data can be estimated by L^2–norms of the boundary data and the right-hand side of the equation. Obviously this additional fact greatly improves our energy estimate. This is achieved by using some intrinsic properties of the area integral and maximal functions.

Finally, I would like to thank to Dr K. Matthews (University of Queensland) for his help in improving the presentation of the material in this book. My sincere thanks to Dr N. Jacob (University Erlangen-Nürnberg) who read the manuscript and whose comments helped me to remove a number of mistakes. Also, I would like to express my gratitude to Professor G. Lieberman (Iowa State University) for allowing me to use his unpublished manuscript.

Brisbane, September 1989.

CHAPTER 1

WEIGHTED SOBOLEV SPACE $\widetilde{W}^{1,2}$

In this chapter we briefly discuss the solvability of the Dirichlet problem for linear elliptic equations in a Sobolev space $W^{1,2}$. The $W^{1,2}$- approach to the Dirichlet problem covers a wide range of boundary data. However, the use of this space does not allow us to solve the Dirichlet problem with every boundary data in L^2. To overcome this difficulty we introduce a larger weighted Sobolev space $\widetilde{W}^{1,2}$. In this chapter we examine properties of the space $\widetilde{W}^{1,2}$ that will be needed to solve the Dirichlet problem with every boundary data in L^2.

1.1. Sobolev spaces.

We recall some fundamental definitions and results from the theory of Sobolev spaces. These spaces are defined over an arbitrary domain $\Omega \subseteqq R_n$. Let $u \in L^1_{\text{loc}}(\Omega)$. A function v_α such that

$$\int_\Omega u(x) D^\alpha \Phi(x)\, dx = (-1)^{|\alpha|} \int_\Omega v_\alpha(x)\Phi(x)\, dx$$

for every $\Phi \in C_\circ^\infty(\Omega)$ is called the weak or distributional partial derivative of u and is denoted by $D^\alpha u$. Here α denotes multi-index $\alpha = (\alpha_1, ..., \alpha_n)$, where α_i are non-negative integers. It is clear that if such a v_α exists, it is unique up to sets of measure zero.

To define Sobolev spaces we introduce a functional $\|.\|_{m,p}$, where m is a non-negative integer and $1 \leqq p \leqq \infty$ as follows:

$$\|u\|_{m,p} = \Big(\sum_{0 \leqq |\alpha| \leqq m} \int_\Omega |D^\alpha u(x)|^p\, dx \Big)^{\frac{1}{p}},$$

for $1 \leqq p < \infty$ and

$$\|u\|_{m,\infty} = \max_{0 \leqq |\alpha| \leqq m} \sup_\Omega |D^\alpha u(x)|$$

if $p = \infty$. It is clear that $\|.\|_{m,p}$ and $\|.\|_{m,\infty}$ define norms on any vector space of functions for which values of these functionals are finite, provided functions are identified in the space if they are equal almost everywhere.

For any integer $1 \leqq m$ and any $1 \leqq p \leqq \infty$ we can define the following spaces:

$$H^{m,p}(\Omega) \equiv \text{the completion of } \{u \in C^m(\Omega);\ \|u\|_{m,p} < \infty\}$$
$$\text{with respect to the norm } \|\cdot\|_{m,p}.$$

$$W^{m,p}(\Omega) \equiv \{u \in L^p(\Omega);\ D^\alpha u \in L^p(\Omega) \quad \text{for} \quad 0 \leqq |\alpha| \leqq m\}.$$

Here D^α denotes the weak partial derivative of u.

$$\overset{\circ}{W}{}^{m,p}(Q) \equiv \text{ the closure of } C_\circ^\infty(\Omega) \text{ in the space } W^{m,p}(\Omega).$$

The above spaces equipped with the norm $\|\cdot\|_{m,p}$ are called Sobolev spaces over Ω. It is clear that $W^{0,p} = \overset{\circ}{W}{}^{0,p} = L^p$. It is also known that $H^{m,p}(\Omega) = W^{m,p}(\Omega)$ for every domain Ω. The spaces $\overset{\circ}{W}{}^{1,2}(\Omega)$ and $W^{1,2}(\Omega)$ are Hilbert spaces with scalar product

$$(u,v) = \int_\Omega (uv + \sum_{i=1}^{n} D_i u D_i v)\, dx.$$

Throughout this work we make frequent use of the Sobolev inequality

$$\|u\|_{L^{2^*}(\Omega)} \leqq S\|Du\|_{L^2(\Omega)},$$

which holds for all $u \in \overset{\circ}{W}{}^{1,2}(\Omega)$, where $\frac{1}{2^*} = \frac{1}{2} - \frac{1}{n}$ and

$$S = (n\pi(n-2))^{-\frac{1}{2}} \left\{ \frac{n!}{2\Gamma(\frac{n}{2})\Gamma(\frac{n}{2}+1)} \right\}^{\frac{1}{n}}.$$

All other properties of Sobolev spaces needed in this work will be mentioned throughout the text.

1.2. The Dirichlet problem.

This work deals with the Dirichlet problem for elliptic equations of the form

$$Lu = -D_i(a^{ij}(x)D_j u + b^i(x)u) + d^i(x)D_i u + c(x)u = f(x) \tag{1.1}$$

in a bounded domain $Q \subset R_n$. The summation convention, that repeated indices indicate summation from 1 to n, is adhered to throughout. The formal adjoint L^* of L is defined by

$$L^*u = -D_i(a^{ji}(x)D_j u + d^i(x)u) + b^i(x)D_i u + c(x)u = f(x).$$

To motivate our approach to the Dirichlet problem, assume that the coefficients of the operator L satisfy the following conditions:

(a) L is uniformly elliptic; that is, there exists a positive constant γ such that

$$\gamma^{-1}|\xi|^2 \leqq a^{ij}(x)\xi_i\xi_j \leqq \gamma|\xi|^2$$

for almost all $x \in Q$ and all $\xi \in R_n$.

(b) $a^{ij} \in L^\infty(Q)\,(i,j = 1,...,n)$, $\quad b^i, d^i \in L^n(Q)\,(i = 1,...n)$ and $c \in L^{\frac{n}{2}}(Q)$.

We associate with L and L^* the following bilinear forms defined on $\overset{\circ}{W}{}^{1,2}(Q) \times \overset{\circ}{W}{}^{1,2}(Q)$:

$$a(u,\varphi) = \int_Q (a^{ij}(x)D_j u D_i\varphi + b^i(x)u D_i\varphi + d^i(x)D_i u\varphi + c(x)u\varphi)\, dx$$

and

$$a^*(u,\varphi) = \int_Q (a^{ji}(x)D_j u D_i\varphi + d^i(x)uD_i\varphi + b^i(x)D_i u\varphi + c(x)u\varphi)\,dx$$

for $u,\varphi \in \overset{\circ}{W}{}^{1,2}(Q)$.

Let $f \in L^2(Q)$. A function u is said to be a weak or generalized solution of the equation (1.1) (or $L^*u = f$) if $u \in W^{1,2}_{\text{loc}}(Q)$ and u satisfies

$$a(u,v) = \int_Q f(x)v(x)\,dx \quad \left(\text{ or } a^*(u,v) = \int_Q f(x)v(x)dx\right)$$

for every $v \in W^{1,2}(Q)$ with a compact support in Q.

A function u is called a solution of the Dirichlet problem

$$Lu = f \text{ in } Q \quad \left(\text{or } L^*u = f\right) \text{ and } u = 0 \text{ on } \partial Q$$

if u is a weak solution in $\overset{\circ}{W}{}^{1,2}(Q)$ of the equation $Lu = f$ (or $L^*u = f$).

There is an extensive literature on the Dirichlet problem. We state here without proof the fundamental results, the proofs of which can be found in [**ST1**], [**ST2**] or [**LU**].

Theorem 1.1. *Suppose that the operator L satisfies the conditions (a) and (b). Then one and only one of the following possibilities holds true:*

(1) *For each $f \in L^2(Q)$ there exists a unique solution $u \in \overset{\circ}{W}{}^{1,2}(Q)$. In particular the homogeneous equation admits only a trivial solution in $\overset{\circ}{W}{}^{1,2}(Q)$.*

(2) *The homogeneous equation $Lu = 0$ has a nontrivial solution in $\overset{\circ}{W}{}^{1,2}(Q)$.*

Furthermore, when (1) holds, the solution operator $A : L^2(Q) \longrightarrow \overset{\circ}{W}{}^{1,2}(Q)$, defined by $Au = f$, is linear and bounded.

In the case (2), the set of solutions of the equation $Lu = 0$ is a finite dimensional subspace of $\overset{\circ}{W}{}^{1,2}(Q)$, and its dimension is the same as the dimension of the set of all solutions in $\overset{\circ}{W}{}^{1,2}(Q)$ of the homogeneous adjoint equation $L^*u = 0$.

Finally, in the case (2), the equation $Lu = f$ has a solution if and only if $\int_Q f(x)v(x)\,dx = 0$ for all solutions $v \in \overset{\circ}{W}{}^{1,2}(Q)$ of the equation $L^*v = 0$.

Together with the equation $Lu = 0$ we consider the corresponding homogeneous equation

$$Lu + \lambda u = 0, \tag{1.2}$$

where λ is a real parameter. If for some λ, there exists a non-trivial solution of (1.2) in $\overset{\circ}{W}{}^{1,2}(Q)$, the λ is called an eigenvalue of (1.2) and the solution is an eigenfunction. The linear space consisting of all the eigenfunctions which correspond to some λ is called the eigenspace corresponding to λ.

Theorem 1.2. *Let the coefficients of L satisfy the assumptions (a) and (b). Then there exists a countable discrete set of eigenvalues $\Sigma \subset (-\infty, \infty)$ such that if $\lambda \notin \Sigma$, then the Dirichlet problem in $\overset{\circ}{W}{}^{1,2}(Q)$ for the equations*

$$Lu + \lambda u = f \quad \text{and} \quad L^*u + \lambda u = f$$

is uniquely solvable for arbitrary $f \in L^2(Q)$.

If $\lambda \in \Sigma$, then the eigenspaces for the equations $Lu + \lambda u = 0$ and $L^*u + \lambda u = 0$ are of the same finite dimension and the Dirichlet problem in $\overset{\circ}{W}{}^{1,2}(Q)$ for the equations $Lu + \lambda u = f$ is solvable if and only if $\int_Q f(x)v(x)\,dx = 0$ for all eigenfunctions v of the equation $L^*v + \lambda v = 0$.

Theorems 1.1 and 1.2 remain true if f is replaced by an element of $W^{-1,2}(Q)$, where $W^{-1,2}(Q)$ denotes the space of linear and continuous functionals on $W^{1,2}(Q)$ i.e.

$$W^{-1,2}(Q) = W^{1,2}(Q)^*.$$

It is well known that every element $T \in W^{-1,2}(Q)$ has the form

$$T(v) = \int_Q (f(x)v(x) + \sum_{i=1}^{n} f_i(x)D_i v(x))\,dx$$

for $v \in W^{1,2}(Q)$, where f and $f_i (i = 1, ..., n)$ are suitable elements in $L^2(Q)$.

This observation allows us to solve the Dirichlet problem with non-zero boundary data φ. Namely, let $\varphi \in W^{1,2}(Q)$. A weak solution u in $W^{1,2}(Q)$ of the equation (1.1) is a solution of the Dirichlet problem with the boundary condition $v = \varphi$ on ∂Q if and only if $u - \varphi \in \overset{\circ}{W}{}^{1,2}$.

One can reduce this problem to the Dirichlet problem with zero boundary data by introducing a new unknown function $v = u - \varphi$. In this situation v satisfies the non-homogeneous equation $Lv = T$, where T is a suitable element in $W^{-1,2}(Q)$.

The above definition of the Dirichlet problem can be reformulated using the concept of a trace of an element in $W^{1,2}(Q)$. The concept of trace can be introduced for domains of class $C^{0,1}$.

Let k be a nonnegative integer and $\alpha \in (0,1)$. A bounded domain Q and its boundary ∂Q are of class $C^{k,\alpha}$, if at each point $x_\circ \in \partial Q$, there exists a ball $B = B(x_\circ)$ and a one-to-one function: $a : B \to D \in R_n$ having the following proper ties : $a(B \cap Q) \subset R^n_+$, $a(B \cap \partial Q) \subset \partial R^n_+$, $a \in C^{k,\alpha}(B)$ and $a^{-1} \in C^{k,\alpha}(D)$.

It is very well known that if ∂Q is of the class $C^{0,1}$ then functions in $W^{1,2}(Q)$ have traces on ∂Q; this means, there exists a linear and compact mapping $T : W^{1,2}(Q) \longrightarrow L^2(\partial Q)$ such that

$$\|Tu\|_{L^2(\partial Q)} \leq C\|u\|_{W^{1,2}(Q)}$$

and if $u \in C^1(\bar{Q})$, then $Tu = u_{|\partial Q}$, where $u_{|\partial Q}$ denotes the restriction of u to ∂Q and C is a positive constant independent of u (see [NE] or [MI1]).

The inclusion $T(W^{1,2}(Q)) \subset L^2(\partial Q)$ is strict.

Let $\varphi \in L^2(\partial Q)$ and assume that there exists $\Phi \in W^{1,2}(Q)$ such that $T\Phi = \varphi$ on ∂Q. A weak solution $u \in W^{1,2}(Q)$ of the equation (1.1) is a solution of the Dirichlet problem with the boundary condition $u = \varphi$ on ∂Q if and only if $u - \Phi \in \overset{\circ}{W}{}^{1,2}(Q)$.

This definition is rather restrictive, because not every function in $L^2(\partial Q)$ is the trace of some function belonging to $W^{1,2}(Q)$. In Lemma 1.1 below, we give an example of a continuous function defined on $\partial B(0,1)$, where $B(0,1)$ is a unit disc in R_2, which is not a trace of any function belonging to $W^{1,2}(Q)$. Using polar coordinates, functions defined on $\partial B(0,1)$ can be viewed as functions defined on the interval $[0,2\pi]$.

Lemma 1.1. *A function $\varphi \in L^2[0,2\pi]$ is the trace of a function belonging to $W^{1,2}(B(0,1))$ if and only if*

$$\sum_{k=1}^{\infty} k(a_k^2 + b_k^2) < \infty,$$

where a_k and b_k denote the Fourier coefficients of φ:

$$a_k = \frac{1}{\pi}\int_0^{2\pi} \varphi(s)\cos ks\, ds \quad k = 0,1,...,$$

$$b_k = \frac{1}{\pi}\int_0^{2\pi} \varphi(s)\sin ks\, ds \quad k = 1,2,...$$

(For the proof see monograph [MI1].)

It follows from the above lemma that the following continuous function on $[0,2\pi]$

$$\sum_{k=1}^{\infty} \frac{\cos k^3 s}{k^2}$$

cannot be extended into function belonging to $W^{1,2}(Q)$.

1.3. Domains with C^2-boundary.

Throughout this chapter Q stands for a bounded domain in R_n with the boundary ∂Q of class C^2. It follows from the regularity of the boundary ∂Q that there exists a number $\delta_\circ$ such that for $\delta \in (0, \delta_\circ]$, the domain

$$Q_\delta = Q \cap \{x;\quad \min_{y\in\partial Q} |x-y| > \delta\}$$

with the boundary ∂Q_δ, possesses the following property: to each $x_\circ \in \partial Q$ there is a unique point $x_\delta(x_\circ) \in \partial Q_\delta$ such that $x_\delta(x_\circ) = x_\circ - \delta\nu(x_\circ)$, where $\nu(x_\circ)$ is the outward normal to ∂Q at $x_\circ$. The inverse mapping of $x_\circ \to x_\delta(x_\circ))$ is given by the formula $x_\circ = x_\delta + \delta\nu_\delta(x_\delta)$, where $\nu_\delta(x_\delta)$ is the outward normal to ∂Q_δ at x_δ.

Let $x_\circ \in \partial Q,\quad 0 < \delta < \delta_\circ$ and let $\bar{x}_\delta$ be given by $\bar{x}_\delta = x_\delta(x_\circ) = x_\circ - \delta\nu(x_\circ)$. For $\epsilon > 0$ we set

$$A_\epsilon = \partial Q_\delta \cap \{x_\delta;\quad |x_\delta - \bar{x}_\delta| < \epsilon\}$$

and

$$B_\epsilon = \{x;\quad x = \tilde{x}_\delta + \delta\nu_\delta(\tilde{x}_\delta),\quad \tilde{x}_\delta \in A_\epsilon\}.$$

Let $\frac{dS_\delta}{dS_\circ} = \lim_{\epsilon\to 0}\frac{|A_\epsilon|}{|B_\epsilon|}$, where $|A|$ denotes the $(n-1)$-dimensional Hausdorff measure of a set A.

Lemma 1.2. *There exists a positive constant $\gamma_\circ$ such that*

$$\gamma_\circ^{-2} \leqq \frac{dS_\delta}{dS_\circ} \leqq \gamma_\circ^2$$

for all $\delta \in (0, \delta_\circ]$ and moreover $\lim_{\delta \to 0} \frac{dS_\delta}{dS_\circ} = 1$ uniformly on ∂Q.

PROOF. It follows from the regularity of ∂Q, that there is a finite sequence of balls $U_1, ..., U_N$ such that

$$\partial Q \subset \bigcup_{i=1}^{N} U_i$$

and for every k, the part $(\partial Q)_k = \partial Q \cap U_k$ of the boundary ∂Q admits a parametrization

$$x = \chi^k(t') \quad (x_i = \chi_i^k(t') \quad i = 1, ..., n)$$

for $t' = (t_1, ..., t_{n-1}) \in D_k$, where D_k is $(n-1)$-dimensional domain with the boundary of class C^2 and moreover $\chi_i^k(t') \in C^2(\bar{D}_k)$ and the vectors $\chi_{t_1}^k, ..., \chi_{t_{n-1}}^k$ are linearly independent on $\bar{D}^k$. Put

$$\Omega_k = \{x : \quad x = x_\circ - \tau\nu(x_\circ), \quad x_\circ \in (\partial Q)_k, \quad 0 < \tau < \delta_\circ\}$$

$k = 1, ..., N$ and introduce new coordinates $(t_1, ..., t_{n-1}, t_n)$ on Ω_k in the following way:

$$x = \chi^k(t') - t_n \nu(\chi^k(t')). \tag{1.3}$$

The vector function $\chi(t') - t_n\nu(\chi^k(t'))$ belongs to $C^1(\{t' \in \bar{D}_k, \quad 0 \leqq t_n \leqq \delta_\circ\})$ and the Jacobian of the transformation (1.3) is given by

$$J_k(t) = -(\chi_{t_1}^k - t_n\nu_{t_1}, ..., \chi_{t_{n-1}}^k - t_n\nu_{t_{n-1}}, \nu),$$

where we have used the following notation: for $a^1 = (a_1^1, ..., a_n^1), ..., a^n = (a_1^n, ..., a_n^n)$. We put $(a^1, ..., a^n) = \det(a_j^i)$.

It is clear that $J_k \neq 0$ on the set $\{t' \in \bar{D}_k, \quad 0 \leqq t_n \leqq \delta_\circ\}$, provided $\delta_\circ$ is sufficiently small. Therefore there exists a positive number $\gamma_\circ$ such that

$$\gamma_\circ^{-1} \leqq J_k \leqq \gamma_\circ \tag{1.4}$$

for all $t' \in \bar{D}_k, \quad 0 \leqq t_n \leqq \delta_\circ, \quad k = 1, ..., N$. Since $Q - \bar{Q}_{\delta_\circ} \subset \bigcup_{k=1}^N \Omega_k$, for every $x_\delta \in \partial Q_\delta \quad (0 < \delta \leqq \delta_\circ)$ there exists $k_\circ \quad (1 \leqq k_\circ \leqq N)$ such that $x_\delta \in \Omega_{k_\circ}$ and so

$$\frac{dS_\delta}{dS_\circ} = \frac{(\chi_{t_1}^{k_\circ} + \delta\nu_{t_1}(t'), ..., \chi_{t_{n-1}}^{k_\circ}(t') + \delta\nu_{t_{n-1}}(t'), \nu(t'))}{(\chi_{t_1}^{k_\circ}(t'), ..., \chi_{t_n}^{k_\circ}(t'), \nu(t'))} = \frac{J_{k_\circ}(t', \delta)}{J_{k_\circ}(t', 0)},$$

where $(t_1, ..., t_{n-1}, \delta)$ and $(t_1, ..., t_{n-1}, 0)$ denote t-coordinates of the points x_δ and $x_\circ$, respectively. Consequently it follows from (1.4) that

$$\gamma_\circ^{-2} \leqq \frac{dS_\delta}{dS_\circ} \leqq \gamma_\circ^2$$

for all $\delta \in (0, \delta_\circ]$ and $\lim_{\delta \to 0} \frac{dS_\delta}{dS_\circ} = 1$ uniformly on ∂Q.

Throughout this work we frequently refer to the following lemma

Lemma 1.3. *Suppose that Q is a bounded domain with the boundary ∂Q of class C^k, $k \geqq 1$. If $f \in C^k(\partial Q)$, then there exists a function $F \in C^k(\bar{Q})$ such that $F(x) = f(x)$ on ∂Q and*

$$\|F\|_{C^k(\bar{Q})} \leqq C\|f\|_{C^k(\partial Q)}$$

for some constant $C > 0$ independent of f.

(For the proof see the monograph [MI1].)

1.4. Some properties of weighted Sobolev spaces.

Let $x \in Q$, and let $r(x)$ denote the distance from x to the boundary ∂Q. Since the boundary ∂Q is of class C^2 the distance $r(x)$ belongs to $C^2(\bar{Q} - Q_{\delta_\circ})$ if $\delta_\circ$ is sufficiently small. Denote by $\rho(x)$ the extension of the function $r(x)$ into $\bar{Q}$ satisfying the following properties: $\rho(x) = r(x)$ for $x \in \bar{Q} - Q_{\delta_\circ}$, $\rho \in C^2(\bar{Q})$, $\rho(x) \geqq \frac{3\delta_\circ}{4}$ in $Q_{\delta_\circ}$, $\gamma_1^{-1} r(x) \leqq \rho(x) \leqq \gamma_1 r(x)$ in Q for some positive constant γ_1, $\partial Q_\delta = \{x;\ \rho(x) = \delta\}$ for $\delta \in (0, \delta_\circ]$ and finally $\partial Q = \{x;\ \rho(x) = 0\}$.

For $u \in W^{1,2}_{\mathrm{loc}}(Q)$ we define the surface integrals

$$M_1(\delta) = \int_{\partial Q} |u(x_\delta(x))|^2\, dS_x \quad \text{and} \quad M(\delta) = \int_{\partial Q_\delta} |u(x)|^2\, dS_x$$

and the values of $u(x_\delta(x))$ on ∂Q and $u(x)$ on ∂Q_δ are understood in the sense of traces. By standard properties of functions in $W^{1,2}_{\mathrm{loc}}$, $M_1(\delta)$ and $M(\delta)$ are absolutely continuous on $[\delta_1, \delta_\circ]$ for every $0 < \delta_1 < \delta_\circ$ (see [NE]).

Lemma 1.4. *If $u \in W^{1,2}_{\mathrm{loc}}(Q)$ and $M(\delta)$ is bounded on $(0, \delta_\circ]$, then for every $0 \leqq \alpha < 1$ there is a positive constant C such that*

$$\int_{Q_\delta} \frac{|u(x)|^2}{(\rho(x) - \delta)^\alpha}\, dx \leqq C,$$

for every $\delta \in (0, \frac{\delta_\circ}{2}]$.

PROOF. Let $\delta \in (0, \frac{\delta_\circ}{2}]$. The result follows from the following estimate

$$\begin{aligned}
\int_{Q_\delta} \frac{u(x)^2}{(\rho(x) - \delta)^\alpha}\, dx &\leqq \int_{Q_{\delta_\circ}} \frac{u(x)^2}{(\rho(x) - \delta)^\alpha}\, dx + \int_{Q_\delta - Q_{\delta_\circ}} \frac{u(x)^2}{(\rho(x) - \delta)^\alpha}\, dx \\
&\leqq \left(\frac{4}{\delta_\circ}\right)^\alpha \int_{Q_{\delta_\circ}} u(x)^2\, dx + \int_\delta^{\delta_\circ} \frac{dt}{(t - \delta)^\alpha} \int_{\partial Q_t} u(x)^2\, dS_x \\
&\leqq \left(\frac{4}{\delta_\circ}\right)^\alpha \int_{Q_{\delta_\circ}} u(x)^2\, dx + \frac{\delta_\circ^{-\alpha+1}}{1 - \alpha} \sup_{0 < \delta \leqq \delta_\circ} M(\delta).
\end{aligned}$$

Lemma 1.5. *Suppose that $u \in W^{1,2}_{\text{loc}}(Q)$ and that $\int_Q |Du(x)|^2 r(x)\, dx < \infty$. Then if $0 \leqq \mu < 1$ and $0 < \delta_1 \leqq \frac{\delta_0}{2}$ we have, for $\delta \in (0, \frac{\delta_1}{2}]$*

$$\int_{Q_\delta} \frac{u(x)^2}{(\rho(x) - \delta)^\mu}\, dx \leqq K\Big(\delta_1^{-\mu} \int_{Q_{\delta_1}} u(x)^2\, dx + \delta_1^{1-\mu} \int_{\partial Q_{\delta_1}} u(x)^2\, dS_x$$

$$+ \delta_1^{1-\mu} \int_{Q_\delta - Q_{\delta_1}} |Du(x)|^2 (\rho(x) - \delta)\, dx\Big),$$

where K is a positive constant inedependent of δ_1 and δ.

PROOF. Let $\delta \in (0, \frac{\delta_1}{2}]$ and put

$$\int_{Q_\delta} \frac{u^2}{(\rho - \delta)^\mu}\, dx = \int_{Q_\delta - Q_{\delta_1}} \frac{u^2}{(\rho - \delta)^\mu}\, dx + \int_{Q_{\delta_1}} \frac{u^2}{(\rho - \delta)^\mu}\, dx.$$

Since $\rho(x) \geqq \delta_1$ on Q_{δ_1}, we have

$$\int_{Q_{\delta_1}} \frac{u^2}{(\rho - \delta)^\mu}\, dx \leqq \left(\frac{2}{\delta_1}\right)^\mu \int_{Q_{\delta_1}} u^2\, dx.$$

We now note that

$$\int_{Q_\delta - Q_{\delta_1}} \frac{u^2}{(\rho - \delta)^{-\mu}}\, dx = \int_\delta^{\delta_1} (t - \delta)^{-\mu} \int_{\partial Q} u(x_t(x_\circ))^2 \frac{dS_t}{dS_\circ}\, dS_\circ dt$$

$$\leqq \gamma_\circ^2 \int_\delta^{\delta_1} (t - \delta)^{-\mu} \int_{\partial Q} u(x_t(x_\circ))^2\, dS_\circ dt.$$

As $\int_{\partial Q} u(x_t(x))^2\, dS_x$ is absolutely continuous on $[\delta, \delta_1]$, integrating by parts gives

$$\int_{Q_\delta - Q_{\delta_1}} \frac{u^2}{(\rho - \delta)^\mu}\, dx \leqq \frac{\gamma_\circ^2 \delta_1^{1-\mu}}{1 - \mu} \int_{\partial Q} u(x_{\delta_1}(x))^2\, dS$$

$$+ \frac{2\gamma_\circ^2}{1 - \mu} \int_\delta^{\delta_1} (t - \delta)^{1-\mu} \int_{\partial Q} |u((x_t(x_\circ))||Du(x_t(x_\circ))||\frac{\partial}{\partial t} x_t(x_\circ)|\, dS_\circ dt$$

$$\leqq \frac{\gamma_\circ^4 \delta_1^{1-\mu}}{1 - \mu} \int_{\partial Q_{\delta_1}} u^2\, dS + \frac{2\gamma_\circ^4}{1 - \mu} \int_{Q_\delta - Q_{\delta_1}} |u(x)||Du(x)|(\rho - \delta)^{1-\mu}\, dx$$

$$\leqq \frac{\gamma_\circ^4 \delta_1^{1-\mu}}{1 - \mu} \int_{\partial Q_{\delta_1}} u^2\, dS + \frac{2\beta\gamma_\circ^4}{1 - \mu} \int_{Q_\delta - Q_{\delta_1}} \frac{u^2}{(\rho - \delta)^\mu}\, dx$$

$$+ \frac{2\gamma_\circ^4 \delta_1^{1-\mu}}{\beta(1 - \mu)} \int_{Q_\delta - Q_{\delta_1}} |Du|^2 (\rho - \delta)\, dx,$$

where we have used Young's inequality in the final step. Now choosing $\frac{2\gamma_\circ^4 \beta}{1-\mu} = \frac{1}{2}$, the result follows.

We now are in a position to define the basic Sobolev space $\widetilde{W}^{1,2}(Q)$ in which we shall study the Dirichlet problem with the boundary data in L^2. This space is defined by

$$\widetilde{W}^{1,2}(Q) = \{u \in W^{1,2}_{\text{loc}}(Q);\ \int_Q u(x)^2\,dx + \int_Q |Du(x)|^2 r(x)\,dx < \infty\}$$

and equipped with the norm

$$\|u\|^2_{\widetilde{W}^{1,2}} = \int_Q u(x)^2\,dx + \int_Q |Du(x)|^2 r(x)\,dx.$$

There is an extensive literature on weighted Sobolev spaces (see monographs [**KU**], [**KS**], and [**NE1**]). In these monographs more general weighted Sobolev spaces are studied, namely,

$$W^{1,2}(Q, r^\epsilon) = \{u \in W^{1,2}_{\text{loc}}(Q),\ \int_Q u(x)^2 r(x)^\epsilon\,dx + \int_Q |Du(x)|^2 r(x)^\epsilon\,dx < \infty\},$$

where ϵ is a fixed real number. Lemma 1.5 shows that $W^{1,2}(Q,r) = \widetilde{W}^{1,2}(Q)$. The proof of this result, under the assumption $\partial Q \in C^{0,1}$, can be found in [**KU**].

In this work we also frequently use the following embedding theorem.

Theorem 1.3. *The space $\widetilde{W}^{1,2}(Q)$ is compactly embedded in $L^2(Q)$.*

For the proof see [**NE1**] or [**ME**].

1.5. Problem of traces in weighted Sobolev spaces.

The result on the existence of traces in $W^{1,2}(Q)$ remains true for the weighted Sobolev space $W^{1,2}(Q, r^\epsilon)$ with $0 \leqq \epsilon < 1$ (see [**NE1**]).

Here we present an elegant argument due to E.T.Poulsen (see [**PO**]) showing the importance of the condition $0 \leqq \epsilon < 1$. Also, this argument leads to an example of a function in $\widetilde{W}^{1,2}(Q)$ which does not have a trace on ∂Q.

We restrict our attention to the integral of the form

$$\int_Q a^{ij}(x) D_i u D_j u\,dx,$$

which plays a crucial role in studying the Dirichlet problem using methods of the theory of Sobolev spaces. In what follows we assume that the coefficients a^{ij} are measurable on Q and that the bilinear form

$$A(x;\xi,\eta) = a^{ij}(x)\xi_i\eta_j$$

is uniformly positive on compact subsets K of Q, that is, there exists a constant $\alpha = \alpha(K) > 0$ such that

$$A(x;\xi,\xi) \geqq \alpha|\xi|^2$$

for all $x \in K$ and all $\xi \in R_n$. Moreover, we assume that the matrix $\{a^{ij}\}$ is symmetric on Q.

Let

$$V = \{u \in W^{1,2}_{\text{loc}}(Q);\ \int_Q a^{ij}(x)D_iuD_ju\,dx < \infty\}$$

and set

$$\|u\|_A^2 = \int_Q a^{ij}(x)D_iuD_ju\,dx.$$

If Q' is a subdomain of Q with the compact closure in Q, then $\int_{Q'} |Du(x)|^2\,dx < \infty$ for all $u \in V$ and hence $u \in L^2(Q')$ for all $u \in V$. For $u \in V$ we define the norm

$$\|u\|_V^2 = \|u\|_A^2 + \|u\|^2_{L^2(Q')}$$

and it is clear that V equipped with this norm becomes a Hilbert space. By V° we denote the closure of $C_\circ^\infty(Q)$ in V. For a domain $\Gamma \subset R_{n-1}$ we define $\Gamma_h = \Gamma \times (x_n = h)$ and $\Gamma_{k,h} = \Gamma \times (k < x_n \leqq h)$. For a point $x \in R_n$ we use the notation $x = (x', x_n)$ with $x' \in R_{n-1}$.

Theorem 1.4. *Let Γ be a bounded domain in R_{n-1} and let $\bar{\Gamma} \times (0 < x_n \leqq a) \subset Q$, $\Gamma_\circ \subset \partial Q$. Moreover, we assume that there exists a function $s(t)$ continuous and positive on $(0, a]$ with $\int_0^a \frac{dt}{s(t)} < \infty$ and such that*

$$A(x;\xi,\xi) \geqq s(x_n)\,|\xi|^2$$

for all $x \in \Gamma_{0,a}$ and all $\xi \in R_n$. Then

(1) *The mapping $h \to u(\cdot, h)$ is continuous from $[0, a]$ into $L^2(\Gamma)$ for all $u \in V$.*
(2) *The mapping $u \to u(\cdot, h)$ is compact from V into $L^2(\Gamma)$ for each $h \in [0, a]$.*
(3) *$u(\cdot, 0) = 0$ for all $u \in V^\circ$.*

PROOF. It is sufficient to show (1) and (2) for $u \in C^\infty(\bar{Q})$. Let $v \in C^\infty(\bar{Q})$ and let $0 < t_1 < t_2 \leqq a$. Then we have

$$v(y,t_2) - v(y,t_1) = \int_{t_1}^{t_2} \frac{\partial v}{\partial x_n}(y,t)\,dt$$

for all $y \in \Gamma$. Hence

$$|v(y,t_2) - v(y,t_1)|^2 \leqq \int_{t_1}^{t_2} \frac{dt}{s(t)} \int_{t_1}^{t_2} s(t)\left|\frac{\partial v}{\partial x_n}\right|^2 dt$$
$$\leqq \int_{t_1}^{t_2} \frac{dt}{s(t)} \int_{t_1}^{t_2} A(y',t;Dv,Dv)\,dt$$

and integrating in y we obtain

$$\|v(\cdot,t_2) - v(\cdot,t_1)\|_{L^2(\Gamma)} \leqq \int_{t_1}^{t_2} \frac{dt}{s(t)} \int_{\Gamma_{t_1,t_2}} A(x;Dv,Dv)\,dx.$$

It is obvious that this inequality holds for all $v \in C^\infty(\bar{Q})$ and consequently

$$\|u(\cdot,t_2) - u(\cdot,t_1)\|_{L^2(\Gamma)} \leqq \int_{t_1}^{t_2} \frac{dt}{s(t)} \|u\|_V^2$$

for all $v \in V$. Since the mapping $u \to u(\cdot,t)$ of V into $L^2(\Gamma)$ is compact for $t > 0$, the last inequality implies (1) and (2). To prove (3) we notice that $u(\cdot,0) = 0$ for all $u \in C_\circ^\infty$ and consequently we have $u(\cdot,0) = 0$ for all $u \in V^\circ$.

The following theorem is converse to Theorem 1.3.

Theorem 1.5. *Let Γ be a bounded domain in R_{n-1} with $\Gamma_{0,a} \subset Q$ and $\Gamma_\circ \subset \partial Q$. Suppose that there exists a function $s(t)$ continuous and positive on $(0,a]$ with $\int_0^a \frac{dt}{s(t)} < \infty$, such that $A(x;\xi,\xi) \geqq s(x_n)|\xi|^2$ for all $x \in \Gamma_{0,a}$ and $\xi \in R_n$, and $a_{nn}(x) \leqq Ks(x_n)$ for all $x \in \Gamma_{0,a}$ and some constant $K > 0$. Finally, suppose that a function $u \in V$ satisfies the following conditions:*

(1) *u has zero boundary values on $\Gamma_\circ$,*
(2) *the support of u is a compact subset of $Q \cup \Gamma_\circ$.*

Then $u \in V^\circ$.

PROOF. To prove this theorem we construct a sequence of functions $v_k = \omega_k u$ in V° such that $\lim_{k\to\infty} v_k = u$ in V. We look for ω_k of the form $\omega_k(x) = f_k(r(x))$ with f_k having the properties: i) f_k satisfies the Lipschitz condition on $[0,\infty)$, ii) $f_k = 0$ for $t \leqq \epsilon_k$, $0 \leqq f_k(t) \leqq 1$ for $\epsilon_k \leqq t \leqq k^{-1}$ and $f_k(t) = 1$ for $k^{-1} \leqq t$. The functions f_k and the numbers ϵ_k will be determined later on. One can easily show that $\omega_k u \in V^\circ$ for all k. Since u has a compact support contained in $Q \cup \Gamma_\circ$ there exists a $k_\circ$ such that

$$v_k(x) = \begin{cases} u(x) & \text{for } x \in Q - \Gamma_{0,k^{-1}} \\ f_k(x_n)u(x) & \text{for } x \in \Gamma_{0,k^{-1}} \end{cases}$$

for $k \geqq k_\circ$. Let $k \geqq k_\circ$, then

$$\|u - v_k\|_V^2 = \int_{\Gamma_{0,k^{-1}}} A(x; D((1-\omega_k)u), D((1-\omega_k)u))\,dx = I_k^1 + I_k^2 + I_k^3,$$

where

$$I_k^1 = \int_{\Gamma_{0,k^{-1}}} (1-\omega_k)^2 A(x; Du, Du)\,dx$$

$$I_k^2 = \int_{\Gamma_{0,k^{-1}}} u^2 A(x; D(1-\omega_k), D(1-\omega_k))\,dx,$$

$$I_k^3 = 2\int_{\Gamma_{0,k^{-1}}} (1-\omega_k)uA(x; Du, D(1-\omega_k))\,dx.$$

By the Cauchy inequality we have

$$|A(x; Du, D(1-\omega_k))|^2 \leqq A(x, Du, Du)A(x, D(1-\omega_k)D(1-\omega_k))$$

and consequently

$$
\begin{aligned}
2|(1-\omega_k)uA(x;Du,D(1-\omega_k))| \\
&\leqq 2\Big[(1-\omega_k)^2A(x;Du,Du)\Big]^{\frac{1}{2}}\Big[u^2A(x;D(1-\omega_k),D(1-\omega_k))\Big]^{\frac{1}{2}} \\
&\leqq (1-\omega_k)^2A(x;Du,Du)+u^2A(x;D(1-\omega_k),D(1-\omega_k)).
\end{aligned}
$$

Therefore $|I_k^3| \leqq |I_k^1|+|I_k^2|$. Thus to prove our theorem we must construct the functions f_k such that $lim_{k\to\infty}I_k^j = 0$ for $j=1,2$. It is obvious that

$$
I_k^1 \leqq \int_{\Gamma_{0,k-1}} A(x;Du,Du)\,dx \to 0 \qquad \text{as } k\to\infty.
$$

For I_k^2 we have

$$
\begin{aligned}
&\int_{\Gamma_{0,k-1}} u^2 a_{nn}(x)|\frac{df_k}{dt}(x_n)|^2\,dx \\
&\leqq \int_{\Gamma_{0,k-1}} u^2Ks(x_n)|\frac{df_k}{dt}|^2\,dx \leqq K\int_0^{k^{-1}} s(t)|\frac{df_k}{dt}|^2\int_\Gamma u(y,t)^2\,dy\,dt \\
&\leqq K\|u\|_V^2\int_0^{k^{-1}} s(t)|\frac{df_k}{dt}|^2\int_0^t\frac{d\tau}{s(\tau)}\,dt.
\end{aligned}
$$

Here we have used the inequality

$$
\|u(\cdot,t_2)-u(\cdot,t_1)\| \leqq \int_0^t\frac{d\tau}{s(\tau)},
$$

with $t_2=t,\quad t_1=0$ and $u(\cdot,0)=0$, which has been established in Theorem 1.3. Let $\Phi(t)=\int_0^t\frac{d\tau}{s(\tau)}$. We now make a choice of f_k, namely,

$$
f_k(t)=\frac{1}{k}\log\frac{\Phi(t)}{\Phi(k^{-1})}+1
$$

with ϵ_k defined as the unique root of the equation $f_k(t)=0$. Taking this expression as f_k on $[\epsilon_k,k^{-1}]$ we get

$$
\int_0^{k^{-1}} s(t)\Phi(t)|\frac{df_k}{dt}|^2\,dt=\int_{\epsilon_k}^{k^{-1}}\frac{k^{-2}}{s(t)\Phi(t)}\,dt=k^{-2}\log\frac{\Phi(k^{-1})}{\Phi(\epsilon_k)}=k^{-1}
$$

and the result easily follows.

The next theorem shows that if $A(x;\xi,\eta)$ strongly degenerates near the boundary ∂Q, then functions in V may not have a trace on ∂Q. Moreover, $V=V^\circ$ if this level of degeneracy holds along ∂Q.

Theorem 1.6. *Let $\Gamma \subset R_{n-1}$ be a bounded domain such that $\bar{\Gamma} \times (0,a] \subset Q$ and $\Gamma_\circ \subset \partial Q$. Suppose that there exists a function $s(t)$ continuous and positive on the interval $(0,a]$ with $\int_0^k \frac{dt}{s(t)} = \infty$ such that*

$$A(x;\xi,\xi) \geqq s(x_n)|\xi|^2$$

for $x \in \Gamma_{0,a}$ and

$$a_{nn}(x) \leqq Ks(x_n)$$

for $x \in \Gamma_{0,a}$. If $u \in V$ and u has a compact support in $Q \cup \Gamma_\circ$, then $u \in V^\circ$.

PROOF. As in the proof of Theorem 1.4 we construct a sequence $v_k = \omega_k u$ of functions in V°. For k sufficiently large we have

$$\|u - v_k\|_V = \int_{\Gamma_{0,k-1}} A(x; D((1-f_k)u), D((1-f_k)u))\,dx \leqq 2(I_k^1 + I_k^2),$$

where

$$I_k^1 = \int_{\Gamma_{0,k-1}} (1-f_k)^2 A(x; Du, Du)\,dx \to 0 \quad \text{as } k \to \infty$$

and

$$I_k^2 = \int_{\Gamma_{0,k-1}} u^2 A(x; D(1-f_k), D(1-f_k))\,dx \to 0 \quad \text{as } k \to \infty.$$

If we set $\Phi(t) = \int_t^a \frac{d\tau}{s(\tau)}$, we get from the proof of Theorem 1.3 that

$$\|u(\cdot,a) - u(\cdot,t)\|_{L^2(\Gamma)} \leqq \Phi(t)\|u\|_V^2$$

and since $\|u(\cdot,a)\|_{L^2(\Gamma)} \leqq C\|u\|_V$ for some $C > 0$, we obtain $\|u(\cdot,t)\|_{L^2(\Gamma)}^2 \leqq C\Phi(t)\|u\|_V$ for all $0 < t \leqq \frac{a}{2}$. Thus for $k \geqq \frac{2}{a}$, sufficiently large, we have

$$I_k^2 = \int_0^{k^{-1}} \int_\Gamma u^2 a_{nn}(x)|\frac{df_k}{dt}(x_n)|^2\,dx'dx_n \leqq KC\|u\|_V^2 \int_0^{k^{-1}} s(t)\Phi(t)|\frac{df_k}{dt}|^2\,dt.$$

Let us now choose

$$f_k(t) = 1 + k^{-1}\log\frac{\Phi(k^{-1})}{\Phi(t)}$$

for $\epsilon_k \leqq t \leqq k^{-1}$, where ϵ_k is determined by the relation $\log\frac{\Phi(\epsilon_k)}{\Phi(k^{-1})} = k$. Consequently we get

$$I_k^2 \leqq KC\|u\|_V^2 k^{-1} \to 0 \quad \text{as } k \to \infty$$

and this completes the proof.

Let us now define the mapping $T_t : V \to L^2(\Gamma)$ by the formula $T_t u = u(\cdot,t)$. It is clear that $T_t u \to 0$ as $t \to 0$ for $u \in C_\circ^\infty(Q) \subset V^\circ$. On the other hand, let $v \in C_\circ^\infty(\Gamma)$, with $v \neq 0$, and define $u(x) = u(x',x_n) = v(x')$ for $0 < x_n < a$ and extend this function to a C^∞-function with compact support in $Q \cup \Gamma_\circ$. In this case we have $\lim_{t\to 0} T_t u = v \neq 0$. Consequently, by the principle of the uniform boundedness, there are functions $u \in V^\circ$ for which $\lim_{t\to 0} T_t u$ does not exist.

CHAPTER 2

THE DIRICHLET PROBLEM IN A HALF–SPACE

In this chapter we solve the Dirichlet problem with L^2–boundary data in a half–space. It is relatively easy to solve this problem under the assumption that the leading coefficients a^{ij} of the operator L have bounded partial derivatives. The aim of this chapter is to show that this assumption can be replaced by a Dini condition. The main ingredient in our approach to the Dirichlet problem is a suitable form of the energy estimate. The energy inequality method allows us to solve the problem of traces of $\widetilde{W}^{1,2}$–solutions and establish the existence theorems for the Dirichlet problem with L^2–boundary data.

2.1. Preliminaries.

As in Section 1.5 we denote a point in R_n by $x = (x', x_n)$ with $x' \in R_{n-1}$ and by R_n^+ the subset of R_n on which x_n is positive. Let l be a fixed positive number, for $x \in R_n^+$ we set $R_n^+(x) = R_n^+ \cap B(x,l)$, where $B(x,l) = \{y;\ |x-y| < l\}$.

For every $1 \leqq p < \infty$, we define the following spaces:

$$M^p(R_n^+) = \{u \in L^p_{\text{loc}}(\bar{R}_n^+);\ \|u\|_{M^p} = \sup_{x\in R_n^+} \|u\|_{p,R_n^+(x)} < \infty\}$$

and

$$M^p_\circ(R_n^+) = \{u \in L^p_{\text{loc}}(\bar{R}_n^+);\ \lim_{|x|\to\infty} \|u\|_{p,R_n^+(x)} = 0\}.$$

It is obvious that $M^p_\circ(R_n^+) \subset M^p(R_n^+)$ and we equip $M_\circ(R_n^+)$ with the norm of the space $M^p(R_n^+)$. Throughout this chapter we assume that $n \geqq 3$. The case $n = 2$ is much easier. We now recall some properties of the spaces M^p and $M^p_\circ$. For the proofs we refer to the papers [**TT**] and [**TR**]. It is very easy to show that $L^p(R_n^+) \subset M^q_\circ(R_n^+)$ for $q \leqq p < \infty$ and moreover $L^\infty(R_n^+) \subset M^p(R_n^+)$ for all $1 \leqq p < \infty$.

Let β be a non–negative and measurable function on R_n^+. We denote by $L^p(R_n^+,\beta)$ the space of functions defined by

$$L^p(R_n^+,\beta) = \{u \in L^1_{\text{loc}}(R_n^+);\ u\beta^{\frac{1}{p}} \in L^p(R_+)\}$$

equipped with the seminorm

$$\|u\|_{L^p(R_n^+,\beta)} = \|\beta^{\frac{1}{p}} u\|_p.$$

If $\beta \in M^{\frac{n}{2}}(R_n^+)$, then $\overset{\circ}{W}{}^{1,2}(R_n^+)$ is continuously imbedded in $L^2(R_n^+,\beta)$, that is, there exists a constant k independent of β and n such that

$$\|u\|_{L^2(R_n^+,\beta)} \leqq k\|\beta\|^{\frac{1}{2}}_{M^{\frac{n}{2}}} \|u\|_{1,2}. \tag{2.1}$$

To prove the inequality (2.1) we write $R_n^+ = \bigcup_{i\in I} \Omega_i$, where I is a countable set, with (i) diam $\Omega_i < d$ for each i, (ii) every Ω_i has a uniform cone property and (iii) there exists a positive integer k such that every intersection of $k+1$ sets from the family $\{\Omega_i\}$ is empty (see [**AD**], p.106). Since each Ω_i satisfies the uniform interior cone property, we have by the Sobolev inequality

$$\int_{\Omega_i} u^2\beta\,dx \leqq \left(\int_{\Omega_i} \beta^{\frac{n}{2}}\,dx\right)^{\frac{2}{n}} \left(\int_{\Omega_i} |u|^{\frac{2n}{n-2}}\,dx\right)^{\frac{n-2}{n}} \leqq C\|\beta\|_{M^{\frac{n}{2}}} \|u\|^2_{W^{1,2}(\Omega_i)}.$$

Now the condition (iii) yields the estimate (2.1). The inequality (2.1) continues to hold for an arbitrary unbounded domain Ω (for details we refer to the paper [**TR**]). If $\beta \in M_o^{\frac{n}{2}}(R_n^+)$, then the imbedding of $\overset{\circ}{W}{}^{1,2}(R_n^+)$ into $L^2(R_n^+,\beta)$ is compact. However we shall not need this result here.

We need the following property of the space $M^p(R_n^+)$.

Lemma 2.1. *Let $f \in M^p(R_n^+)$. Then for each $0 < q < p$ and $\epsilon > 0$ f admits a decomposition $f = f_1 + f_2$, with $\|f_1\|_{M^q} < \epsilon$ and $f_2 \in L^\infty(R_n^+)$.*

PROOF. For $R > 0$ we write

$$f(x) = f(x)\chi_{\{|f(x)|>R\}} + f(x)\chi_{\{|f(x)|\leqq R\}} = f_1(x) + f_2(x),$$

where for $A \subset R_n^+$ χ_A denotes the characteristic function of a set A. For $x \in R_n^+$ we have

$$\begin{aligned}\|f\|_{M^p} \geqq \|f_1\|_{M^p} &\geqq \left(\int_{R_n^+(x)} |f_1(y)|^p\,dy\right)^{\frac{1}{p}} \\ &= \left(\int_{B(x,l)\cap\{|f(y)|>R\}} |f(u)|^p\,dy\right)^{\frac{1}{p}} \geqq R|B(x,l)\cap\{|f(y)|>R\}|.\end{aligned}$$

This inequality implies that

$$\lim_{R\to\infty} |B(x,l)\cap\{|f(y)|>R\}| = 0$$

uniformly in $x \in R_n^+$. To complete the proof we observe that for $q < p$ we have

$$\begin{aligned}\left(\int_{B(x,l)} f_1(y)|^q\,dy\right)^{\frac{1}{q}} &= \left(\int_{B(x,l)\cap\{|f(y)|>R\}} |f(y)|^q\,dy\right)^{\frac{1}{q}} \\ &\leqq \left(\int_{B(x,l)} |f(y)|^p\,dy\right)^{\frac{1}{p}} |B(x,l)\cap\{|f(y)|>R\}|^{\frac{1}{q}-\frac{1}{p}}\end{aligned}$$

and the result follows by taking R sufficiently large.

2.2. Assumptions and properties of solutions in $W^{1,2}_{\text{loc}}(R_n^+)$.

In R_n^+ we consider the operator L defined by (1.1). We make the following assumptions on the coefficients of L.

(A) The coefficients a^{ij} $(i,j=1,...,n)$ are in $L^\infty(R_n^+)$ and

$$a^{ij}(x)\xi_i\xi_j \geqq \gamma|\xi|^2$$

for all $x \in R_n^+$ and some constant $\gamma > 0$, $a^{ij} = a^{ji}$, $(i,j=1,..,n)$.

(B) There is a bounded concave nondecreasing function $\omega \in C^1(0,\infty)$ with

$$\int_0^1 t^{-1}\omega(t)^2\,dt < \infty \tag{2.2}$$

such that

$$|a^{ni}(x) - a^{ni}(x',0)| \leqq \omega(x_n)$$

for all $x \in R_n^+$ and $i = 1,...,n$,

$$|a^{ni}(x',0) - a^{ni}(x_1,...,x_{i-1},y,x_{i+1},...,x_{n-1},0)| \leqq \omega(|x_i - y|)^2$$

for all $x' \in R_{n-1}$, $y \in (-\infty,\infty)$, and $i = 1,...,n-1$.

We investigate the existence of a solution to the Dirichlet problem under two different sets of assumptions on the coefficients $b = (b^1,...,b^n)$, $d = (d^1,...,d^n)$ and c.

(C_1) The coefficients b^i, d^i and c admit the following decompositions

$$b + d = h + \bar{h}, \quad b = k + \bar{k}, \quad c = c_1 + c_2 \tag{2.3}$$

with

$$\frac{h}{\omega(q)} \in M^{n+\delta_1}(R_n^+), \quad \frac{k'q}{\omega(q)^2} \in M^{n+\delta_2}(R_n^+), \quad \frac{k^n}{\omega(q)^2} \in M^{n+\delta_3}(R_n^+),$$
$$\frac{c_1}{\omega(q)^2} \in M^{\frac{n}{2}+\delta_4}(R_n^+)$$

for some positive constants δ_1, δ_2, δ_3 and δ_4, and

$$\frac{\bar{h}q}{\omega(q)}, \frac{\bar{k}'q^2}{\omega(q)^2}, \frac{\bar{k}^n q}{\omega(q)^2}, \frac{c_2q^2}{\omega(q)^2} \in L^\infty(R_n^+) \tag{2.4}$$

where $q = \min\{1, x_n\}$.

(C_2) The coefficients b^i, d^i and c admit the decompositions (2.3) with

$$\frac{h}{\omega(q)} \in M_o^n(R_n^+), \quad \frac{k'q}{\omega(q)^2} \in M_o^n(R_n^+), \quad \frac{k^n}{\omega(q)^2} \in M_o^n(R_n^+),$$
$$\frac{c_1}{\omega(q)^2} \in M_o^{\frac{n}{2}}(R_n^+),$$

and $\bar{h}$, $\bar{k}$ and c_2 satisfying (2.4).

We commence with some properties of solutions in $W^{1,2}_{\text{loc}}(R_n^+)$ of the equation

$$Lu = -D_i g^i + f \text{ in } R_n^+. \tag{2.5}$$

We make the following assumptions on the functions g^i and f

(D) $$\frac{fq^{\frac{3}{2}}}{\omega(q)},\ g'q^{\frac{1}{2}},\ \frac{g^n q^{\frac{1}{2}}}{\omega(q)} \in L^2(R_n^+).$$

We note here that $u \in W^{1,2}_{\text{loc}}(R_n^+)$ is a weak solution of the equation (2.5) if

(2.6) $$a(u,v) = \int_{R_n^+} (g^i D_i v + fv)\, dx$$

for all $v \in W^{1,2}(R_n^+)$ with compact support in R_n^+.

Next we define

$$W_*^{1,2}(R_n^+) = \{u \in W^{1,2}_{\text{loc}}(R_n^+);\ \int_{\{x_n>T\}} u(x)^2\, dx < \infty \text{ for all } T>0\}.$$

In Lemmas 2.2 and 2.3, below, we use the following version of the trace estimate, namely, there exists a constant $C>0$ such that for any $u \in W^{1,2}(R_n^+)$ we have

(2.7) $$\int_{R_{n-1}} u(x',\delta)^2\, dx' \leqq C\|u\|^2_{W^{1,2}}$$

for all $\delta > 0$.

Lemma 2.2. *Let $u \in W_*^{1,2}(R_n^+)$ be a solution of (2.5). Suppose that the assumption (C_1) holds. Then there exists a constant $M>0$ such that*

(2.7) $$\sup_{\{2r<\delta\}} \int_{R_{n-1}} u(x',\delta)^2\, dx + \int_{\{x_n>2r\}} |Du(x)|^2\, dx \leqq M \int_{\{x_n>r\}} [u^2 + f^2 + |g|^2]\, dx.$$

PROOF. Let $v = u\Phi^2$, where $\Phi \in C_\circ^\infty(R_n)$ with $\operatorname{supp} \Phi \subset \{x;\ x' \in R_{n-1}, x_n > r\}$. Using v as a test function we obtain

$$\int_{R_n^+} a^{ij} D_i u D_j u \Phi^2\, dx + 2\int_{R_n^+} a^{ij} D_i u u D_j \Phi\Phi\, dx + \int_{R_n^+} (b^i + d^i) D_i u u \Phi^2\, dx + 2\int_{R_n^+} u^2 b^i D_i \Phi\Phi\, dx + \int_{R_n^+} cu^2\Phi^2\, dx = \int_{R_n^+} fu\Phi^2\, dx + \int_{R_n^+} g^i D_i u \Phi^2\, dx + 2\int_{R_n^+} u g^i D_i \Phi\Phi\, dx.$$

The integrals involving the coefficients b^i, d^i and c can be estimated using Lemma 2.1 and the Sobolev inequality. We restrict ourselves to the third integral on the left hand side. Using the decomposition $b^i + d^i = h^i + \bar{h}^i$ we obtain for $\epsilon > 0$

$$\left|\int_{R_n^+} \bar{h}^i D_i u u \Phi^2\, dx\right| \leqq \epsilon \int_{R_n^+} |Du|^2\Phi^2\, dx + C(\epsilon)\int_{R_n^+} u^2\Phi^2\, dx,$$

where $C(\epsilon) = \epsilon^{-1}|||\bar{h}|||_\infty$. To estimate the integral involving h^i we use the decomposition from Lemma 2.1 $h^i = e_1^i + e_2^i$ with $\|e_1^i\|_{M^n} < \epsilon$ and $e_2^i \in L^\infty(R_n^+)$ $(i = 1, ..., n)$. Applying the estimate (2.1) we get

$$|\int_{R_n^+} h^i D_i uu\Phi^2\, dx| \leqq k|||e_1|^2\|_{M^{\frac{n}{2}}}^{\frac{1}{2}} \|u\Phi\|_{1,2}\|Du\Phi\|_2$$
$$+ \epsilon \int_{R_n^+} |Du|^2\Phi^2\, dx + C_1(\epsilon) \int_{R_n^+} u^2\Phi^2\, dx$$
$$\leqq k\epsilon^{\frac{1}{2}}\left[2\int_{R_n^+} |Du|^2\Phi^2\, dx + \int_{R_n^+} u^2(\Phi^2 + |D\Phi|^2)\, dx\right]$$
$$+ \epsilon \int_{R_n^+} |Du|^2\Phi^2\, dx + C_1(\epsilon) \int_{R_n^+} u^2\Phi^2\, dx,$$

where $C_1(\epsilon) = |||e_2|||_\infty \epsilon^{-1}$. Combining these estimates together we obtain

$$\left|\int_{R_n^+} (b^i + d^i) D_i uu\Phi^2\, dx\right| \leqq (2\epsilon + 2\epsilon^{\frac{1}{2}} k) \int_{R_n^+} |Du|^2\Phi^2\, dx$$
$$+ C_2(\epsilon) \int_{R_n^+} u^2(\Phi^2 + |D\Phi|^2)\, dx$$

for some positive constant $C_2(\epsilon)$. In a similar way we estimate the fourth and fifth integral on the left hand side. Consequently applying the Young inequality and the ellipticity of L and choosing ϵ sufficiently small, we arrive at the estimate

$$\int_{R_n^+} |Du|^2\Phi^2\, dx \leqq C\left[\int_{R_n^+} u^2(\Phi^2 + |D\Phi|^2)\, dx + \int_{R_n^+} f^2\Phi^2\, dx\right.$$
$$\left. + \int_{R_n^+} |g|^2\Phi^2\, dx\right],$$

where a constant $C > 0$ depends on the norms of the coefficients on $\{x_n > r\}$, n and γ. Now let $\{\Phi_\nu\}$ be an increasing sequence of non–negative functions in $C_\circ^\infty(R_n^+)$ with the gradients bounded independently of ν and converging to a non–negative function on R_n^+ equal to 1 for $x_n \geqq 2r$ and vanishing for $x_n < r$. If we substitute $\Phi = \Phi_\nu$ and let ν tend to infinity we obtain the estimate for $\int_{\{x_n > r\}} |Du|^2\, dx$. It follows from this estimate that $u \in W^{1,2}(R_{n-1} \times [r, \infty))$ and the application of the trace estimate (2.7) completes the proof.

To prove Lemma 2.3 we use, in place of Lemma 2.1, the following property of the space $M_\circ^p(R_n^+)$. Namely, let $f_a \in C_\circ^\infty(R_n)$ be such that $f_a(x) = 1$ on $B(0, a)$, $0 \leqq f_a \leqq 1$ on R_n and supp $f_a \subset B(0, 2a)$. If $u \in M_\circ^p(R_n^+)$ then

$$\lim_{a\to\infty} \|(1 - f_a)u\|_{M^p} = 0. \tag{2.9}$$

This is an easy consequence of the definition of the space $M_\circ^p$.

Lemma 2.3. *Suppose that the assumption* (C_2) *holds. Then the estimate (2.8) continues to hold for any solution* $u \in W_*^{1,2}(R_n^+)$ *of (2.5).*

PROOF. We follow the argument used in the proof of Lemma 2.2. We illustrate the use of the property (2.9) by estimating the integral involving $b^i + d^i$. We write this integral in the form

$$\int_{R_n^+} (b^i + d^i) D_i u u \Phi^2 \, dx = \int_{R_n^+} h^i (1 - f_a) D_i u u \Phi^2 \, dx$$
$$+ \int_{R_n^+} h^i f_a D_i u u \Phi^2 \, dx + \int_{R_n^+} \bar{h}^i D_i u u \Phi^2 \, dx$$
$$= i_1 + i_2 + i_3.$$

It follows from (2.1) and (2.9) that for a given $\epsilon > 0$ we have

$$|i_1| \leqq k \|(1 - f_a)^2 |h|^2\|_{M^{\frac{n}{2}}}^{\frac{1}{2}} \|u\Phi\|_{1,2} \|Du\Phi\|_2$$
$$\leqq k\epsilon \left[2 \int_{R_n^+} |Du|^2 \Phi^2 \, dx + \int_{R_n^+} u^2 |D\Phi|^2 \, dx \right]$$

provided we choose a sufficiently large. To estimate i_2 we observe that $h^i f_a \in L^n(R_n^+)$. Consequently we have the decomposition

$$h^i f_a = h^i f_a \chi_{\{|h^i f_a| > R\}} + h^i f_a \chi_{\{|h^i f_a| < R\}} = s^i + \bar{s}^i$$

with $\|s^i\|_n < \epsilon$ for R sufficiently large and $\bar{s}^i \in L^\infty(R_n^+)$, $(i = 1, ..., n)$. Using the Young and Sobolev inequalities we obtain

$$|i_2| \leqq \||s|\|_n \|u\Phi\|_{2^*} \|Du\Phi\|_2 + \||\bar{s}|\|_\infty \int_{R_n^+} |Du||u|\Phi^2 \, dx$$
$$\leqq \epsilon S \left[2 \int_{R_n^+} |Du|^2 \Phi^2 \, dx + \int_{R_n^+} u^2 |D\Phi|^2 \, dx \right] + \epsilon \int_{R_n^+} |Du|^2 \Phi^2 \, dx$$
$$+ C_3(\epsilon) \int_{R_n^+} u^2 \Phi^2 \, dx,$$

where $C_3(\epsilon) = \||\bar{s}|\|_\infty \epsilon^{-1}$. Finally, for i_3 we have

$$|i_3| \leqq \epsilon \int_{R_n^+} |Du|^2 \Phi^2 \, dx + C_4(\epsilon) \int_{R_n^+} u^2 \Phi^2 \, dx,$$

where $C_4(\epsilon) = \||\bar{h}|\|_\infty \epsilon^{-1}$. We complete the proof by applying the approximation argument of Lemma 2.2.

We point out here that the assumption (B) concerning the regularity of the coefficients a^{ni}, is not needed for the validity of Lemmas 2.2 and 2.3. This assumption will play an important role in the next sections.

2.3. Behaviour of $W_*^{1,2}$–solutions for small x_n.

In this section we examine the integrals $\int_{R_n^+} |Du(x)|^2 \min(1, x_n) \, dx$ and $\sup_{0<\delta<d} \int_{R_{n-1}} u(x', \delta) \, dx'$. We need a technical lemma.

Lemma 2.4. *There exist functions A^i $(i = 1, ..., n-1)$ such that*

$$|A^i(x) - a^{ni}(x)| \leqq (1 + \sup \omega)\omega(x_n), \quad |D_i A^i(x)| \leqq 2\frac{\omega(x_n)^2}{x_n}$$

for each $i < n$ and all $x \in R_n^+$.

PROOF. Let η be a non–negative $C^{0,1}(R)$–function with support in the interval $(-1, 1)$ such that $\int_{-\infty}^{\infty} \eta(t)\,dt = 1$ and define

$$A^i(x) = \int_{-\infty}^{\infty} a^{ni}(x^*(i,t))\eta(t)\,dt,$$

where $x^*(i,t) = (x_1, ..., x_{i-1}, x_i - x_n t, x_{i+1}, ..., x_{n-1}, 0)$. The first inequality follows from the estimate

$$|A^i(x) - a^{ni}(x)| \leqq \int_{-\infty}^{\infty} |a^{ni}(x^*(i,t)) - a^{ni}(x', 0)|\eta(t)\,dt$$
$$+ |a^{ni}(x) - a^{ni}(x', 0)| \leqq \int_{-\infty}^{\infty} \omega(x_n|t|)^2\eta(t)\,dt + \omega(x_n) \leqq \omega(x_n)^2 + \omega(x_n).$$

Next we observe that $\int_{-\infty}^{\infty} \eta'(t)\,dt = 0$, and therefore

$$|D_i A^i| = |\frac{1}{x_n}\int_{-\infty}^{\infty} [a^{ni}(x^*(i,t)) - a^{ni}(x', 0)]\eta'(t)\,dt|$$
$$\leqq \frac{\omega(x_n)^2}{x_n}\int_{-\infty}^{\infty} |\eta'(t)|\,dt.$$

The proof is completed by choosing $\eta(t) = \frac{3}{4}\max(1 - t^2, 0)$.

We are now in a position to prove the theorem which plays a crucial role in the treatment of traces on $x_n = 0$.

Theorem 2.1. *Suppose that (C_1) holds and let $u \in W_*^{1,2}(R_n^+)$ be a solution of (2.4). Then the following conditions are equivalent*

(I) $\sup_{0<\delta<1} \int_{R_{n-1}} u(x', \delta)^2\,dx < \infty$,

(II) $\int_{R_n^+} |Du(x)|^2 \min(1, x_n)\,dx < \infty$

PROOF. We commence by showing that (I) $\Rightarrow$ (II). Let $\delta \in (0, 1)$ and define

$$\eta(t) = \max(q^{\frac{1}{2}} - \delta^{\frac{1}{2}}, 0)^2.$$

For $\Phi \in C_\circ^1(R_{n-1})$, we set

$$v(x) = u(x)\eta(x_n)\Phi(x')^2$$

and observe that by Lemma 2.2 v is a legitimate test function in (2.5) and on substitution we obtain

$$\int_{R^*} a^{ij} D_i u D_j u \eta \Phi^2\, dx + \int_{R^*} a^{nj} D_j u u \eta' \Phi^2\, dx + 2\int_{R^*} a^{ij} D_j u u D_i \Phi \Phi \eta\, dx$$
$$+ \int_{R^*} (b^i + d^i) D_i u u \eta \Phi^2\, dx + 2\int_{R^*} u^2 b^i D_i \Phi \Phi \eta\, dx + \int_{R^*} u^2 b^n \eta' \Phi^2\, dx$$
$$+ \int_{R^*} c u^2 \eta \Phi^2\, dx = \int_{R^*} f u \eta \Phi^2\, dx + \int_{R^*} g^i D_i u \eta \Phi^2\, dx$$
$$+ 2\int_{R^*} u g^i D_i \Phi \Phi \eta\, dx + \int_{R^*} u g^n \eta' \Phi^2\, dx,$$

where $R^* = \{x;\ x \in R_n^+ \text{ and } x_n > \delta\}$. We now proceed to estimate the integrals in this identity, which we denote by $J_1, \ldots, J_{11}$. It follows from the assumption (A) that

$$J_1 \geqq \int_{R^*} |Du|^2 \eta \Phi^2\, dx.$$

Setting

$$J_2 = \int_{R^*} a^{nn} D_n u u \eta' \Phi^2\, dx + \sum_{i=1}^{n-1} \int_{R^*} a^{ni} D_i u u \eta' \Phi^2\, dx = J_1' + J_2',$$

we then have

$$J_1' = \frac{1}{2}\int_{R^*} a^{nn}(x', 0) D_n(u^2) \eta' \Phi^2\, dx + \int_{R^*} [a^{nn}(x) - a^{nn}(x', 0)] D_n u u \eta' \Phi^2\, dx$$
$$= I_1 + I_2.$$

Integrating by parts we obtain

$$I_1 = -\frac{1}{2}\int_{R^*} a^{nn}(x', 0) u(x)^2 \eta'' \Phi(x')^2\, dx$$
$$+ \frac{1}{2}\int_{R_{n-1}} a^{nn}(x', 0)[u(x', 1)^2 \eta'(1-) - u(x', \delta)^2 \eta'(\delta+)] \Phi^2\, dx'$$
$$= -\frac{1}{2}\int_{R^*} a^{nn}(x', 0) u(x)^2 \eta''(x_n) \Phi(x')^2\, dx'$$
$$+ \frac{1}{2}(1 - \delta^{\frac{1}{2}}) \int_{R_{n-1}} a^{nn}(x', 0) u(x', 1)^2 \Phi(x')^2\, dx'.$$

Therefore,

$$|I_1| \leqq C_1 \left[\int_{R_{n-1}} u(x', 1)^2 \Phi(x')^2\, dx' + \int_{R^*} u^2 \eta'' \Phi^2\, dx \right],$$

where $C_1 = \|a^{nn}\|_\infty$. To estimate I_2, we note that

$$\frac{|\eta'(t)|^2}{\eta(t)} = \frac{1}{t} \text{ for } \delta < t < 1.$$

We then obtain from (B) and Young's inequality that

$$|I_2| \leqq \int_{R^*} \omega(x_n)\eta'(t)|D_n u||u|\Phi^2\,dx$$
$$\leqq \frac{\gamma}{4}\int_{R^*} |Du|^2\eta\Phi^2\,dx + \frac{4}{\gamma}\int_{R^*}\frac{\omega(q)^2}{q}u^2\Phi^2\,dx.$$

Writing $a^{ni} = [a^{ni} - A^i] + A^i$ in J_2', integrating the integral involving A^i by parts, and using Lemma 2.4, we obtain

$$|J_2'| \leqq \frac{\gamma}{4}\int_{R^*} |Du|^2\eta\Phi^2\,dx + C_2\int_{R^*}\left[\frac{\omega(q)^2}{q} + 1\right]u^2\,dx,$$

where C_2 depends only on $\omega(1)$, $\|a^{ni}\|_\infty$ for $i < n$, γ and $\|\Phi + |D\Phi|\|_\infty$. Using the Cauchy–Schwarz inequality we have

$$|J_3| \leqq \frac{\gamma}{4}\int_{R^*} |Du|^2\eta\Phi^2\,dx + C_3(\gamma)\int_{R^*} u^2|D\Phi|^2\,dx.$$

Let $\epsilon > 0$ be a given number which will be determined later. It follows from the Young inequality that

$$|J_4| \leqq \epsilon\int_{R^*} |Du|^2\Phi^2\eta\,dx + \frac{1}{\epsilon}\int_{R^*} |h|^2u^2\eta\Phi^2\,dx$$
$$+ \left\|\bar{h}\frac{q}{\omega}\right\|_\infty\left[\epsilon\int_{R^*} |Du|^2 q\Phi^2\,dx + \frac{1}{\epsilon}\int_{R^*} u^2\frac{\omega^2}{q}\Phi^2\,dx\right].$$

We now observe that $|h|^2\frac{1}{\omega^2} \in M^{\frac{n}{2}+\frac{\delta_1}{2}}(R_n^+)$ and applying Lemma 2.2 with $p = \frac{n}{2} + \frac{\delta_1}{2}$ and $q = \frac{n}{2}$ we obtain a decomposition for a given ϵ

$$(h^i)^2\frac{1}{\omega^2} = h_1^i + h_2^i$$

with $\||h_1|\|_{M^{\frac{n}{2}}} < \epsilon$ and $h_2^i \in L^\infty(R_n^+)$. Using (2.1) we arrive at the estimate

$$|J_4| \leqq \epsilon\int_{R^*} |Du|^2\Phi^2\eta\,dx + \epsilon k^2\int_{R^*} |D(u\omega\eta^{\frac{1}{2}}\Phi)|^2\,dx$$
$$+ \frac{1}{\epsilon}\||h_2|\|_\infty\int_{R^*} u^2\omega^2\Phi^2\,dx + \left\|\bar{h}\frac{q}{\omega(q)}\right\|_\infty\left[\epsilon\int_{R^*} |Du|^2 q\Phi^2\,dx + \frac{1}{\epsilon}\int_{R^*} u^2\frac{\omega^2}{q}\Phi^2\,dx\right].$$

The concavity of ω implies that $t\omega'(t) \leqq \omega(t)$ for $t \in (0,1)$ and hence

$$|D(\eta^{\frac{1}{2}}\omega(q)| \leqq \frac{3}{2}q^{-\frac{1}{2}}\omega(q)$$

and this gives the estimate for J_4

$$|J_4| \leqq \epsilon\Big(1 + k^2 \sup_{0<t<1} \omega(t) + \Big\| |\bar{h}| \frac{q}{\omega} \Big\|_\infty\Big) \int_{R^*} |Du|^2 q \Phi^2 \, dx$$

$$+ \frac{1}{\epsilon} \||h_2|\|_\infty \int_{R^*} u^2 \omega^2 \eta \Phi^2 \, dx + (\frac{1}{\epsilon} \Big\| |h| \frac{q}{\omega(q)} \Big\|_\infty + \frac{3}{2} \epsilon k^2) \int_{R^*} u^2 \frac{\omega^2}{q} \Phi^2 \, dx$$

$$+ \epsilon k^2 \int_{R^*} u^2 \omega^2 \eta |D\Phi|^2 \, dx.$$

To estimate J_5 we first observe that

$$|J_5| \leqq 2 \int_{R^*} |k'| u^2 |D\Phi| \Phi \eta \, dx + 2 \int_{R^*} |\bar{k}'| u^2 |D\Phi| \Phi \eta \, dx$$

$$\leqq \int_{R^*} \frac{|k'|^2 q^2}{\omega(q)^4} (u \omega q^{\frac{1}{2}} \Phi)^2 + \int_{R^*} \frac{\omega(q)^2}{q} u^2 |D\Phi|^2 \, dx$$

$$+ 2 \Big\| \frac{|\bar{k}'| q^2}{\omega(q)^2} \Big\|_\infty \int_{R^*} \frac{\omega(q)^2}{q} u^2 |D\Phi| \Phi \, dx.$$

It follows from Lemma 2.1 that

$$\frac{qk'}{\omega(q)^2} = k_1' + k_2',$$

with $\||k_1'|\|_{M^n} < \epsilon$ and $k_2' \in L^\infty(R_n^+)$. Using the estimate (2.1) we obtain

$$|J_5| \leqq \epsilon k^2 \int_{R^*} |D(u \omega q^{\frac{1}{2}} \Phi)|^2 \, dx + \||k_2'|\|_\infty \int_{R^*} u^2 \omega^2 q \Phi^2 \, dx$$

$$+ \int_{R^*} \frac{\omega(q)^2}{q} u^2 |D\Phi|^2 \, dx + 2 \Big\| \frac{|\bar{k}'| q^2}{\omega(q)^2} \Big\|_\infty \int_{R^*} \frac{\omega(q)^2}{q} u^2 |D\Phi| \Phi \, dx.$$

Similarly, applying Lemma 2.1, we have

$$\frac{k^n}{\omega^2} = k_1^n + k_2^n, \quad \frac{c_2}{\omega^2} = \bar{c}_1 + \bar{c}_2$$

with $\|k_1^n\|_{M^n} < \epsilon$, $k_2^n \in L^\infty(R_n^+)$ and $\|\bar{c}_1\|_{M^{\frac{n}{2}}} < \epsilon$, $\bar{c}_1 \in L^\infty(R_n^+)$. Hence

$$|J_6 + J_7| \leqq \int_{R^*} (k_1^n)^2 (u \omega(q) q^{\frac{1}{2}} \Phi)^2 \, dx + \int_{R^*} \frac{\omega(q)^2}{q} u^2 \Phi^2 \, dx$$

$$+ \|k_2^n\|_\infty \int_{R^*} u^2 \omega(q)^2 \Phi^2 \, dx + \Big\| \frac{\bar{k}^n q}{\omega(q)^2} \Big\|_\infty \int_{R^*} \frac{\omega(q)^2}{q} \Phi^2 u^2 \, dx$$

$$+ \int_{R^*} \bar{c}_1 (u \eta^{\frac{1}{2}} \omega(q) \Phi)^2 + \|\bar{c}_1\|_\infty \int_{R^*} u^2 \eta \omega(q)^2 \Phi^2 \, dx$$

$$+ \Big\| \frac{c_2 q^2}{\omega(q)^2} \Big\|_\infty \int_{R^*} \frac{\omega(q)^2}{q} u^2 \Phi^2 \, dx \leqq 2\epsilon k^2 \int_{R^*} |D(u \eta^{\frac{1}{2}} \omega \Phi)|^2 \, dx$$

$$+ \Big(1 + \Big\| \frac{\bar{k}^n q}{\omega(q)^2} \Big\|_\infty + \Big\| \frac{c_2 q^2}{\omega(q)^2} \Big\|_\infty\Big) \int_{R^*} \frac{\omega(q)^2}{q} u^2 \Phi^2 \, dx$$

$$+ \|k_2^n\|_\infty \int_{R^*} u^2 \omega(q)^2 \Phi^2 \, dx + \|\bar{c}_1\|_\infty \int_{R^*} u^2 \eta \omega(q)^2 \Phi^2 \, dx.$$

Combining the estimates for J_i and choosing ϵ so that

$$\epsilon\left(1+\left\||\bar{h}|\frac{q}{\omega(q)}\right\|_\infty + k^2 \sup\omega + 4k^2\right) < \frac{\gamma}{4}$$

we arrive at the estimate

$$\int_{R^*} |Du|^2\Phi^2\,dx \leqq K\Big[\int_{R_{n-1}} u(x',1)^2\Phi\,dx + \int_{R^*} \frac{\omega(q)^2}{q} u^2(\Phi^2+|D\Phi|^2)\,dx$$
$$+\int_{R^*} u^2(\Phi^2+|D\Phi|^2)\,dx + \int_{R^*} u^2|\eta''|\Phi^2\,dx + \int_{R^*} f^2\frac{q^3}{\omega(q)^2}\Phi^2\,dx$$
$$+\int_{R^*} |g'|^2 q\Phi^2\,dx + \int_{R^*} (g^n)^2\frac{q}{\omega(q)^2}\,dx,$$

where $K>0$ is a constant depending on the norm of the coefficients n and γ. If (I) holds, then the integrals $\int_{R^*} u^2\,dx$, $\int_{R^*} \frac{\omega(q)^2}{q}u^2\,dx$ and $\int_{R^*} u^2|\eta''|\,dx$ are bounded independently of δ. We now replace Φ by a sequence of non–negative $C_\circ^\infty(R_{n-1})$–functions, with bounded gradients, which increases to 1 pointwise and the condition (II) follows by letting $\delta \to 0$.
To prove (II) $\Rightarrow$ (I), we let μ be a non–negative increasing $C^2([0,\infty))$–function such that $\mu(t)=t$ for $t \leqq \frac{2}{3}$ and $\mu(t)=1$ for $t \geqq 1$. For $\delta \in (0,\frac{1}{2})$ we define

$$v(x) = u(x)\max(\mu(x_n)-\delta,0)\Phi(x')^2,$$

where Φ is as in the step (I) $\Rightarrow$ (II). From Lemma 2.2 we infer that v is a legitimate test function and, therefore, we conclude that

$$\frac{1}{2}\int_{R_{n-1}} a^{nn}(x',0)u(x',\delta)^2\Phi(x')^2\,dx'$$
$$= \int_{R\#} [a^{nn}(x)-a^{nn}(x',0)]D_n uu\mu'\Phi^2\,dx - \frac{1}{2}\int_{R\#} a^{nn}(x',0)u^2\mu''\Phi^2\,dx$$
$$+\sum_{i=1}^{n-1}\int_{R\#} [a^{ni}-A^i]D_i uu\mu'\Phi^2\,dx + \sum_{i=1}^{n-1}\int_{R\#} A^i D_i uu\mu'\Phi^2\,dx$$
$$+\int_{R^*} a^{ij}D_iuD_ju(\mu-\delta)\Phi^2\,dx + 2\int_{R^*} a^{ij}D_iuu(\mu-\delta)D_j\Phi\Phi\,dx$$
$$+\int_{R^*} (d^i+b^i)D_iuu(\mu-\delta)\Phi^2\,dx$$
$$+2\int_{R^*} u^2b^iD_i\Phi\Phi(\mu-\delta)\,dx + \int_{R\#} u^2b^n\mu'\Phi^2\,dx + \int_{R^*} cu^2(\mu-\delta)\Phi^2\,dx$$
$$-\int_{R^*} fu(\mu-\delta)\Phi^2\,dx - \int_{R^*} g^iD_iu\Phi^2(\mu-\delta)\,dx - 2\int_{R^*} g^iD_i\Phi\Phi(\mu-\delta)u\,dx$$
$$-\int_{R^*} g^n\mu'\Phi^2 u\,dx,$$

where $R^{\#} = \{x \in R_n^+;\ \delta < x_n < 1\}$. As in the first part of the proof we show that

$$\int_{R_{n-1}} u(x',\delta)^2 \Phi(x')^2\, dx' \leqq C\Big[\int_{R^*} |Du|^2 q\Phi^2\, dx \tag{2.8}$$
$$+ \int_{R^*} \frac{\omega(q)^2}{q} u^2(\Phi^2 + |D\Phi|^2)\, dx + \int_{R^*} u^2(\Phi^2 + |D\Phi|^2)\, dx$$
$$+ \int_{R^*} f^2 \frac{q^3}{\omega(q)^2}\Phi^2\, dx + \int_{R^*} |g'|^2 q\Phi^2\, dx + \int_{R^*} (g^n)^2 \frac{q}{\omega(q)^2}\, dx\Big].$$

Applying the approximation argument from (I) $\Rightarrow$ (II) we show that this estimate remains true with $\Phi \equiv 1$ on R_{n-1}. Let us now observe that

$$\int_{R^*} (\frac{\omega(q)^2}{q} + 1)u^2\, dx \leqq \int_\delta^d (\frac{\omega(s)^2}{s} + 1)\, ds \int_{R_{n-1}} u(x',s)^2\, dx'$$
$$+ (\frac{\sup \omega^2}{d} + 1)\int_{x_n > d} u(x)^2\, dx \leqq \frac{1}{2} \sup_{\delta < s < d} \int_{R_{n-1}} u(x',s)^2\, dx'$$
$$+ (\frac{\sup \omega^2}{d} + 1)\int_{x_n > d} u(x)^2\, dx$$

for d sufficiently small. If we now set $I(t) = \int_{R_{n-1}} u(x',t)^2\, dx'$ we obtain from (2.10)

$$I(\delta) \leqq \frac{1}{2} \sup_{\delta < t < d} I(t) + C_1$$

if $\delta < \min(d, \frac{1}{3})$ with $C_1 > 0$ independent of δ, and the theorem is proved.

Theorem 2.2. *Suppose that (C_2) holds and let $u \in W_*^{1,2}(R_n^+)$ be a solution of (2.4). Then the conditions (I) and (II) are equivalent.*

The proof is similar to that of Theorem 2.1. To estimate the integrals involving the coefficients b^i, d^i and c we use the property (2.9) of spaces $M_\circ^p$. Lemma 2.3 is needed to justify the legitimacy of the corresponding test functions, which are the same as in the proof of Theorem 2.1.

2.4. Traces of solutions in $W_*^{1,2}(R_n^+)$ on $x_n = 0$.

Let us assume that $u \in W_*^{1,2}(R_n^+)$ is a solution of (2.5) and suppose that one of the conditions (I) or (II) holds. Since bounded sets in $L^2(R_{n-1})$ are weakly compact, there exist $\varphi \in L^2(R_{n-1})$ and a sequence $\delta_\nu \to 0$ as $\nu \to \infty$ such that

$$\lim_{\nu \to \infty} \int_{R_{n-1}} u(x',\delta_\nu)v(x')dx' = \int_{R_{n-1}} \varphi(x')v(x')\, dx'$$

for each $v \in L^2(R_{n-1})$. We now improve this observation by showing that the limit exists as $\delta \to 0$. In the next step we show that $\lim_{\delta \to 0} u(x',\delta) = \varphi(x')$ in $L^2(R_{n-1})$. However we must introduce some assumptions on d^i and more restrictive assumptions on k^i. These additional assumptions come from the nature of test functions that will be used in the proofs.

Theorem 2.3. *Suppose that one of the assumptions* (C_1) *or* (C_2) *holds and that* $d^i = s^i + \bar{s}^i$ *with* $\frac{s^i}{\omega(q)} \in M^n(R_n^+)$ *and* $\frac{\bar{s}^i q}{\omega} \in L^\infty$ $(i = 1, \ldots, n)$. *If* $u \in W_*^{1,2}(R_n^+)$ *is a solution of (2.5) satisfying one of the conditions (I) and (II), then there exists a function* $\varphi \in L^2(R_{n-1})$ *such that*

$$\lim_{\delta \to 0} \int_{R_{n-1}} u(x', \delta) v(x')\, dx' = \int_{R_{n-1}} \varphi(x') v(x')\, dx'$$

for each $v \in L^2(R_{n-1})$.

PROOF. Let μ be the function introduced in the proof of Theorem 2.1 and let $m(t)$ be a smooth function on R such that $m(t) = 1$ for $t \leqq 1$ and $m(t) = 0$ for t large. Taking

$$w(x) = v(x') \max\{\mu(x_n) - \delta, 0\} m(x_n)$$

as a test function, where $v \in C_\circ^\infty(R_{n-1})$, we obtain

$$\int_{R_{n-1}} a^{nn}(x', 0) u(x', \delta) v(x') dx' = \int_{R^*} a^{ij} D_i u D_j v(\mu - \delta) m\, dx$$

$$+ \int_{R^*} a^{nj} D_j u v(\mu - \delta) m'\, dx - \int_{R^*} a^{nn}(x', 0) u v \mu'' m\, dx$$

$$- \int_{R^*} a^{nn}(x', 0) u v \mu' m'\, dx + \int_{R+} [a^{nn}(x) - a^{nn}(x', 0)] D_n u v \mu' m\, dx$$

$$+ \int_{R^*} \sum_{i=1}^{n-1} A_i D_i v \mu' m\, dx + \int_{R^*} \sum_{i=1}^{n-1} [a^{ni} - A^i] D_i u v \mu' m\, dx$$

$$+ \int_{R^*} b^i u D_i v(\mu - \delta) m\, dx + \int_{R^*} b^n u v \mu' m\, dx + \int_{R^*} b^n u v(\mu - \delta) m'\, dx$$

$$+ \int_{R^*} d^i D_i u v(\mu - \delta) m\, dx + \int_{R^*} c u v(\mu - \delta) m\, dx - \int_{R^*} f v(\mu - \delta) m\, dx$$

$$- \int_{R^*} g^i D_i v(\mu - \delta) m\, dx - \int_{R^*} g^n v \mu' m\, dx - \int_{R^*} g^n v(\mu - \delta) m'\, dx.$$

We now show that all the integrals on the right hand side are convergent as $\delta \to 0$. We restrict ourselves only to the integral involving d^i. Using the decomposition $d^i = s^i + \bar{s}^i$ we have

$$\int_{R^*} d^i D_i u v(\mu - \delta) m\, dx = \int_{R^*} s^i D_i u v(\mu - \delta) m\, dx + \int_{R^*} \bar{s}^i D_i u v(\mu - \delta) m\, dx$$
$$= i_1 + i_2.$$

It follows from (2.1) that

$$|i_1| \leqq \int_{R^*} \frac{|s|^2}{\omega(q)^2} \left(v\omega(q) m \frac{(\mu - \delta)}{q^{\frac{1}{2}}} \right)^2 dx + \int_{R^*} |Du|^2 q\, dx$$

$$\leqq k^2 \left\| \frac{|s|^2}{\omega(q)^2} \right\|_{M^{\frac{n}{2}}} \int_{R^*} |D(v\omega(q)(\mu - \delta)^{\frac{1}{2}} m)|^2\, dx + \int_{R^*} |Du|^2 q\, dx,$$

which shows that $\lim_{\delta\to 0} i_1$ exists. To estimate i_2 we use the Young inequality

$$|i_2| \leqq \left\|\frac{|\bar{s}|q}{\omega(q)}\right\|_\infty \left[\int_{R^*} |Du|^2 q\,dx + \int_{R^*} \frac{\omega(q)^2}{q} v^2 m\,dx\right]$$

and again $\lim_{\delta\to 0} i_2$ exists. Since $\int_{R_{n-1}} u(x',\delta)v(x')\,dx'$ is continuous on $[\delta_1, \delta_\circ]$ for each $\delta_1 < \delta_\circ$, the above argument shows that this integral is continuous on $[0, \delta_\circ]$ for each $v \in C_\circ^\infty(R_{n-1})$. Approximating a function $v \in L^2(R_{n-1})$ by a sequence $\{v_\nu\}$ in $C_\circ^\infty(R_{n-1})$ converging to v in $L^2(R_{n-1})$, we show, using (I), that $\int_{R_{n-1}} u(x',\delta)v(x')\,dx'$ is continuous on $[0, \delta_\circ]$ as the limit of uniformly convergent sequence of continuous functions on $[0, \delta_\circ]$.

We are now in a position to establish that $u(\cdot, \delta) \to \varphi$ in $L^2(R_{n-1})$ as $\delta \to 0$. To do so, we first show the convergence of the L^2–norms; the uniform convexity of $L^2(R_{n-1})$ then implies the convergence of $u(\cdot, \delta)$ in $L^2(R_{n-1})$. To establish this result we must modify the assumption on the functions k_i. We first observe that Theorem 2.1 remains true if either $\frac{k}{\omega^2} \in M^{n+\delta_2}(R_n^+)$ for some $\delta_2 > 0$ or $\frac{k}{\omega^2} \in M_\circ^n(R_n^+)$. This is only needed to estimate the integral J_5 in the proof of this theorem. These two assumptions imply that either $\frac{kq}{\omega^2} \in M^{n+\delta_2}(R_n^+)$ or $\frac{kq}{\omega^2} \in M_\circ^n(R_n^+)$, respectively, and this is what is needed in the proof of the next theorem.

Theorem 2.4. *Suppose that the assumptions of Theorem 2.3 hold with the following modification of the assumption on* k*: either* $\frac{k}{\omega^2} \in M^{n+\delta_2}(R_n^+)$ *for some* $\delta_2 > 0$ *or* $\frac{k}{\omega^2} \in M_\circ^n(R_n^+)$. *If* $u \in W_*^{1,2}(R_n^+)$ *is a solution of (2.5) and one of the conditions (I) or (II) holds, then there exists a function* $\varphi \in L^2(R_{n-1})$ *such that*

$$\lim_{\delta\to 0} \int_{R_{n-1}} [u(x',\delta) - \varphi(x')]^2\,dx' = 0.$$

PROOF. For $\Psi \in W^{1,2}(R_n^+)$ we have as in the proof of Theorem 2.3

$$\begin{aligned}
&\int_{R_{n-1}} a^{nn}(x',0)\varphi(x')\Psi(x',0))\,dx' \\
&= \int_{R_n^+} [a^{nn}(x) - a^{nn}(x',0)] D_n u(x)\Psi(x)\mu'(x_n)\,dx \\
&- \int_{R_n^+} D_n(\Psi(x)\mu'(x_n))a^{nn}(x',0)u(x)\,dx \\
&+ \sum_{i=1}^{n-1} \int_{R_n^+} [a^{ni} - A^i] D_i u\mu'\Psi\,dx + \int_{R_n^+} \sum_{i=1}^{n-1} A^i D_i u\mu'\Psi\,dx \\
&+ \int_{R_n^+} a^{ij} D_i u D_j \Psi\mu\,dx + \int_{R_n^+} d^i D_i u\mu\Psi\,dx \\
&+ \int_{R_n^+} b^i D_i \Psi u\mu\,dx + \int_{R_n^+} b^n\mu'\Psi u\,dx + \int_{R_n^+} cu\mu\Psi\,dx \\
&- \int_{R_n^+} f\mu\Psi\,dx - \int_{R_n^+} g^i D_i\Psi\mu\,dx - \int_{R_n^+} g^n\Psi\mu'\,dx.
\end{aligned}$$

We define the quantity $F(\Psi)$ by setting the right member of this identity equal to $\int_{R_n^*} F(\Psi)\,dx$. If we now set

$$x^\# = (x', \frac{1}{2}[x_n + \max(x_n, \delta)]) \text{ and } v_\delta(x) = u(x^\#),$$

it is clear that $v_\delta \in W^{1,2}(R_n^+))$ and therefore,

$$\int_{R_{n-1}} a^{nn}(x',0)\varphi(x')u(x',\frac{\delta}{2})\,dx' = \int_{R_n^+} F(v_\delta)\,dx$$

$$= \int_{x_n<\delta} F(u(x^\#))\,dx + \int_{x_n>\delta} F(u)\,dx = H_1 + H_2.$$

We shall show that

$$\lim_{\delta\to 0} H_2 = \lim_{\delta\to 0} \int_{R_{n-1}} a^{nn}(x',0)u(x',\delta)^2\,dx' \text{ and } \lim_{\delta\to 0} H_1 = 0.$$

Indeed, using

$$v(x) = u(x)\max(\mu(x_n) - \delta, 0),$$

we derive from the integral identity defining the weak solution u that

$$\lim_{\delta\to 0} H_2 = \lim_{\delta\to 0}\Big[-\int_{R^*} a^{nn}(x',0)D_n u(x)u(x)\mu'(x_n)\,dx$$

$$-\int_{R^*} u(D_n u\mu')a^{nn}(x',0)\,dx\Big]$$

$$= \lim_{\delta\to 0}\int_{R_{n-1}} a^{nn}(x',0)u(x',\delta)^2\,dx'.$$

To show that H_1 tends to zero, we now write $R^\delta = \{x \in R_n^+;\ x_n < \delta\}$ and we then have

$$H_1 = \int_{R^\delta} [a^{nn}(x) - a^{nn}(x',0)]D_n u(x)\mu'(x_n)u(x^\#)\,dx$$

$$- \int_{R^\delta} a^{nn}(x',0)u(x)D_n\left(u(x^\#)\mu'(x_n)\right)dx$$

$$+ \sum_{i=1}^{n-1}\int_{R^\delta} [a^{ni}(x) - A^i(x)]D_i u(x)\mu'(x_n)u(x^\#)\,dx$$

$$+ \int_{R^\delta}\sum_{i=1}^{n-1} A^i(x)D_i u(x)u(x^\#)\mu'(x_n)\,dx + \int_{R^\delta} a^{ij}(x)D_i u(x)D_j u(x^\#)\mu(x_n)\,dx$$

$$+ \int_{R^\delta} b^i(x)D_i u(x^\#)u(x)\mu(x_n)\,dx + \int_{R^\delta} b^n(x)u(x^\#)u(x)\mu'(x_n)\,dx$$

$$+ \int_{R^\delta} d^i(x)D_i u(x)u(x^\#)\mu(x_n)\,dx + \int_{R^\delta} c(x)u(x^\#)u(x)\mu(x_n)\,dx$$

$$- \int_{R^\delta} f(x)u(x^\#)\mu(x_n)\,dx - \int_{R^\delta} g^i(x)D_i u(x^\#)\mu(x_n)\,dx$$

$$- \int_{R^\delta} g^n(x)u(x^\#)\mu'(x_n)\,dx.$$

We denote by $I_1, ..., I_{12}$ the integrals on the right–hand side of this equation. Using conditions (I), (II), Lemma 2.4 and (2.1) one can show that each I_j tends to zero as $\delta \to 0$ for $j = 1, ..., 12$. To demonstrate, we restrict our consideration to the integrals I_4, I_6 and I_7. Integrating by parts and applying Lemma 2.4 gives

$$
\begin{aligned}
|I_4| &\leqq \sum_{i=1}^{n-1} \int_{R^\delta} |D_i A^i(x)| |\mu'(x_n)| |u(x)| |u(x^\#)| \, dx \\
&+ \sum_{i=1}^{n-1} \int_{R^\delta} |A^i(x)| |u(x)| |D_i u(x^\#)| |\mu'(x_n)| \, dx \\
&\leqq C \left[\int_{R^\delta} \frac{\omega(q)^2}{q} |u(x)| |u(x^\#)| \, dx + \int_{R^\delta} |u(x)| |Du(x^\#)| \, dx \right] \\
&= C[R_1 + R_2]
\end{aligned}
$$

for some constant $C > 0$. According to (I) R_1 tends to zero while Hölder's inequality implies that

$$
\begin{aligned}
|R_2| &\leqq \left[\int_{R^\delta} \frac{2u(x)^2}{x_n + \delta} \, dx \right]^{\frac{1}{2}} \left[\int_{R^\delta} |Du(x^\#)|^2 \frac{x_n + \delta}{2} \, dx \right]^{\frac{1}{2}} \\
&\leqq \left[2 \sup_{0<s<\delta} \int_{R_{n-1}} u(x', s)^2 \, dx' \int_0^\delta \frac{dt}{t + \delta} \right]^{\frac{1}{2}} \left[\int_{0<x_n<\delta} |Du(x^\#)|^2 \frac{x_n + \delta}{2} \, dx \right]^{\frac{1}{2}} \\
&= \left[\log 2 \sup_{0<t<\delta} \int_{R_{n-1}} u(x', t)^2 \, dx \right]^{\frac{1}{2}} \left[\int_{\frac{1}{2}\delta<x_n<\delta} |Du(x)|^2 x_n \, dx \right]^{\frac{1}{2}}
\end{aligned}
$$

and, consequently, R_2 tends to zero. Hence I_4 tend to zero. For I_6 we have

$$
I_6 = \int_{R^\delta} k^i D_i u(x^\#) u(x) \mu(x_n) \, dx + \int_{R^\delta} \bar{k}^i D_i u(x^\#) u(x) \mu(x_n) \, dx = i_1 + i_2.
$$

It follows from (2.1) that

$$
\begin{aligned}
&|i_1| \\
&\leqq \left[\int_{R^\delta} \frac{|k|^2}{\omega(q)^2} (u(x)\omega(q)\mu^{\frac{1}{2}}(x_n))^2 \frac{2\mu(x_n)}{x_n + \delta} \, dx \right]^{\frac{1}{2}} \left[\int_{R^\delta} |Du(x^\#)|^2 \frac{x_n + \delta}{2} \, dx \right]^{\frac{1}{2}} \\
&\leqq C \left\| \frac{|k|^2}{\omega(q)^2} \right\|_{M^{\frac{n}{2}}} \left[\int_{R^*} |D(u(x)\omega(q)\mu(x_n)^{\frac{1}{2}})|^2 \, dx \right]^{\frac{1}{2}} \left[\int_{R^\delta} |Du(x^\#)|^2 \frac{x_n + \delta}{2} \, dx \right]^{\frac{1}{2}}
\end{aligned}
$$

for some $C > 0$. The integral over R^* is bounded independently of δ, while the integral involving $Du(x^\#)$ converges to zero as $\delta \to 0$. Hence $i_1 \to 0$ as $\delta \to 0$. To estimate i_2 we use the Young inequality to obtain

$$
|i_2| \leqq \left\| \frac{|\bar{k}| q}{\omega(q)^2} \right\|_\infty \left[\int_{R^\delta} |Du(x^\#)|^2 \frac{x_n + \delta}{2} \, dx + 2 \sup \omega^2 \int_{R^\delta} \frac{\omega(q)^2}{q} u(x)^2 \frac{\mu(x_n)}{x_n + \delta} \, dx \right]
$$

It is easy to see that $\lim_{\delta\to 0} i_2 = 0$. To estimate I_7 we again use (2.1) and write a decomposition

$$\begin{aligned} I_7 &= \int_{R^\delta} k^n u(x)u(x^\#)\mu'(x_n)\,dx + \int_{R^\delta} \bar{k}^n u(x)u(x^\#)\mu'(x_n)\,dx \\ &= j_1 + j_2 \end{aligned}$$

and then

$$\begin{aligned} &|j_1| \\ &\leq \left[\int_{R^\delta} \left(\frac{k^n}{\omega(q)^2}\right)^2 (u(x)q^{\frac{1}{2}}\omega(q))^2\,dx\right]^{\frac{1}{2}} \left[\int_{R^\delta} \frac{\omega(q)^2}{q} u(x^\#)^2\,dx\right]^{\frac{1}{2}} \\ &\leq k\left\|\left(\frac{k^n}{\omega(q)^2}\right)^2\right\|_{M^{\frac{n}{2}}}^{\frac{1}{2}} \left[\int_{R^*} |D(u(x)q^{\frac{1}{2}}\omega(q))|^2\,dx\right]^{\frac{1}{2}} \left[\int_{R^\delta} \frac{\omega(q)^2}{q} u(x^\#)^2\,dx\right]^{\frac{1}{2}} \end{aligned}$$

and

$$|j_2| \leq \left\|\frac{\bar{k}^n q}{\omega(q)^2}\right\|_\infty \left[\int_{R^\delta} \frac{\omega(q)^2}{q} u(x)^2\,dx + \int_{R^\delta} \frac{\omega(q)^2}{q} u(x^\#)^2\,dx\right].$$

The last two estimates imply that $\lim_{\delta\to 0} I_7 = 0$.

2.5. The Dirichlet problem in $\widetilde{W}^{1,2}(R_n^+)$.

According to Thoerem 2.4, a $W_*^{1,2}$–solution of (2.5) satisfying one of the conditions I or II has boundary values in $L^2(R_{n-1})$. Therefore, Theorem 2.4 suggests the following approach to the Dirichlet problem.

Let $\varphi \in L^2(R_{n-1})$. A weak solution $u \in W_*^{1,2}(R_n^+)$ of the equation (2.5) is a solution of the Dirichlet problem with the boundary condition

$$u(x',0) = \phi(x') \text{ on } R_{n-1},$$

if

$$\lim_{\delta\to 0} \int_{R_{n-1}} [u(x',\delta) - \varphi(x')]^2\,dx' = 0. \tag{2.11}$$

As it stands, the Dirichlet problem need not have a solution, however in what follows, we shall prove that the modified Dirichlet problem

$$Lu + \lambda u = -D_i g^i + f \text{ in } R_n^+, \quad u(x',0) = \varphi(x') \text{ on } R_{n-1} \tag{2.12}$$

has a unique solution in $W_*^{1,2}(R_n^+)$ provided the real parameter λ is sufficiently large. The existence result is based on the energy estimate.

Theorem 2.5. *Suppose that the assumptions of Theorem 2.4 hold. Then there exist constants $C > 0$, $\lambda_\circ > 0$ and $d > 0$ such that any solution $u \in W_*^{1,2}(R_n^+)$ of the problem (2.12), for $\lambda \geqq \lambda_\circ$, satisfies*

$$\int_{R_n^+} |Du(x)|^2 q\, dx + \sup_{0<\delta<d} \int_{R_{n-1}} u(x',\delta)^2\, dx + (\lambda - \lambda_\circ) \int_{R_n^+} u(x)^2 q\, dx \leqq C\left[\int_{R_{n-1}} \varphi(x')^2\, dx' + F\right], \tag{2.13}$$

where

$$F = \left\| \frac{f q^{\frac{3}{2}}}{\omega(q)} \right\|_2 + \|g' q^{\frac{1}{2}}\|_2 + \left\| g^n \frac{q^{\frac{1}{2}}}{\omega(q)} \right\|_2.$$

PROOF. We first note that each integral on the left–hand side of this inequality is finite because (2.10) implies condition (I) of Theorem 2.1. Next we observe that the proof of $(I) \Rightarrow (II)$ in Theorem 2.1 can be repeated with $\eta = q$ and $\Phi \equiv 1$; the finiteness of the integrals in (2.13) guarantees that $v(x) = u(x)q(x_n)$ is a legitimate test function. It therefore follows that

$$\int_{R_n^+} a^{ij} D_i u D_j u q\, dx + \int_{R^*} a^{nj} D_j u u\, dx + \int_{R_n^+} (b^i + d^i) D_i u u q\, dx + \int_{R^*} b^n u^2\, dx + \int_{R_n^+} c u^2\, dx + \lambda \int_{R_n^+} u^2 q\, dx = \int_{R_n^+} f u q\, dx + \int_{R_n^+} g^i D_i u q\, dx + \int_{R^*} g^n u\, dx,$$

where here R^* is a subset of R_n^+ on which $0 < x_n < 1$. We now write $J_1, \ldots. J_9$ for these integrals, which are estimated as in the proof of $(I) \Rightarrow (II)$ from Theorem 2.1. In particular, we decompose J_2 as follows

$$J_2 = \int_{R^*} a^{nn} D_n u u\, dx + \sum_{j=1}^{n-1} \int_{R^*} a^{nj} D_j u u\, dx = J_2' + J_2'',$$

$$J_2' = \frac{1}{2} \int_{R^*} D_n(a^{nn}(x',0) u(x)^2)\, dx + \int_{R^*} [a^{nn}(x) - a^{nn}(x',0)] D_n u u\, dx = \frac{I_1}{2} + I_2.$$

Integrating by parts we obtain

$$I_1 = \int_{R_{n-1}} a^{nn}(x',0) u(x',1)^2\, dx - \int_{R_{n-1}} a^{nn}(x',0) \varphi(x')^2\, dx'.$$

The remaining integrals are estimated as in Theorem 2.1 to conclude that

$$\int_{R_n^+} |Du|^2 q\,dx + \lambda \int_{R_n^+} u^2 q\,dx \leqq C_1[F + \int_{R_{n-1}} \varphi(x')^2\,dx']$$
$$+ C_2 \int_{R_{n-1}} u(x',1)^2\,dx' + C_3 \int_{R_n^+} (1 + \frac{\omega(q)^2}{q}) u^2\,dx,$$

where the constants C_1 depend on L. Applying Lemma 2.2, in case of the assumption (C_1), and Lemma 2.3, if (C_2) holds, we obtain

$$\int_{R_n^+} |Du|^2 q\,dx + \lambda \int_{R_n^+} u^2 q\,dx \leqq C_4[F + \int_{R_{n-1}} \varphi(x')^2\,dx']$$
$$+ C_5(\delta_\circ) \int_{R_n^+} u^2 q\,dx + C_6(\delta_\circ) \sup_{0<\delta<\delta_\circ} \int_{R_{n-1}} u(x',\delta)^2\,dx',$$

where C_4 depends on L, C_5 and C_6 depends also on $\delta_\circ$, and C_6 tends to zero as $\delta_\circ$ tends to zero. On the other hand from the proof of $(II) \Rightarrow (I)$ we deduce that there are constants δ_1 and C_7 depending on L such that

$$\sup_{0<\delta<\delta_\circ} \int_{R_{n-1}} u(x',\delta)^2\,dx \leqq C_7[F + \int_{R_n^+} |Du|^2 q\,dx + \lambda \int_{R_n^+} u^2 q\,dx]$$

provided $\delta_\circ < \delta_1$. Combining the last two estimates yields

$$\int_{R_n^+} |Du|^2 q\,dx + \sup_{0<\delta<d} \int_{R_n^+} u(x',\delta)^2\,dx' + \lambda \int_{R_n^+} u^2 q\,dx$$
$$\leqq (C_4+1)(C_7+1)[F + \int_{R_{n-1}} \varphi(x')^2\,dx'] + (C_7+1)C_5(\delta_\circ) \int_{R_n^+} u^2 q\,dx$$
$$+ (C_7+1)C_6(\delta_\circ) \sup_{0<\delta<\delta_\circ} \int_{R_{n-1}} u(x',\delta)^2\,dx',$$

provided $\delta_\circ < \min(\delta_1, \frac{1}{3})$. The proof is completed by first choosing $d \leqq \min(\delta_1, \frac{1}{3})$ so small that $(C_7+1)C_\circ(d) \leqq \frac{1}{2}$ and then setting $C = 2(C_4+1)(C_7+1)$ and $\lambda_\circ = 2(C_7+1)C_5(d)$.

We also require the following form of the energy estimate for solutions in $\overset{\circ}{W}{}^{1,2}(R_n^+)$.

Lemma 2.5. *Suppose that either*

$$d^i + b^i \in M^{n+\delta_1}(R_n^+)\,(i = 1,...,n), \quad c \in M^{\frac{n}{2}+\delta_2}(R_n^+)$$

for some $\delta_1 > 0$ and $\delta_2 > 0$, or

$$d^i + b^i \in M_\circ^n(R_n^+)\,(i = 1,...,n) \quad c \in M_\circ^{\frac{n}{2}}(R_n^+).$$

Moreover, we assume that f and g^i ($i = 1, ..., n$) are in $L^2(R_n^+)$. Then there exist constants $C > 0$ and $\lambda_* > 0$ depending on L such that any solution $u \in \overset{\circ}{W}{}^{1,2}(R_n^+)$ of the problem (2.12) (with $\varphi \equiv 0$) for $\lambda \geqq \lambda_*$ satisfies

$$(2.14) \qquad \int_{R_n^+} |Du(x)|^2 \, dx + (\lambda - \lambda_*) \int_{R_n^+} u(x)^2 \, dx \leqq C \left[\int_{R_n^+} f(x)^2 \, dx + \int_{R_n^+} |g(x)|^2 \, dx \right].$$

To prove estimate (2.14) we take as a test function u and employ the method of the proof of Theorem 2.4.

Motivated by Theorem 2.5 we introduce the weighted Sobolev space

$$\widetilde{W}^{1,2}(R_n^+) = \{u \in W_{loc}^{1,2}(R_n^+); \int_{R_n^+} |Du(x)|^2 q(x_n) \, dx + \int_{R_n^+} u(x)^2 \, dx < \infty\}$$

equipped with the norm

$$\|u\|^2_{\widetilde{W}^{1,2}} = \int_{R_n^+} |Du(x)|^2 q(x_n) \, dx + \int_{R_n^+} u(x)^2 \, dx.$$

Theorem 2.6. *Suppose that the assumptions of Theorem 2.4 hold. Let $\varphi \in L^2(R_n^+)$ and let d, C and $\lambda_\circ$ be the constants from Theorem 2.4. If $\lambda > \lambda_\circ$, then the problem (2.12) has a unique solution $u \in \widetilde{W}^{1,2}(R_n^+)$. Moreover, this solution obeys the estimate (2.13).*

PROOF. The estimate (2.13) is guaranteed by Theorem 2.4. We proceed in four steps by proving the existence result under successively weaker assumptions. First we assume that the coefficients and functions f and g^i satisfy the assumptions of Lemma 2.5 and that $\varphi \equiv 0$. In that case the estimate (2.14) and a standard application of the Riesz representation theorem show that the problem (2.12) has a unique solution $u \in \overset{\circ}{W}{}^{1,2}(R_n^+)$.

Next we assume that $\varphi \equiv 0$ and that b^i, d^i ($i = 1, ..., n$), c, f and g satisfy the conditions of Theorem 2.4. We now define the truncation operator T_m for a positive integer m to be the multiplication by characteristic function of the subset of R_n^+ on which $x_n > \frac{1}{m}$. Then, from the first step of this proof, there is a unique solution $u_m \in \overset{\circ}{W}{}^{1,2}(R_n^+)$ of

$$Lu_m + \lambda u_m = -D_i(T_m g^i) + T_m f \text{ in } R_n^+, \; u_m(x', 0) = 0 \text{ on } R_{n-1}$$

provided $\lambda > \lambda_*$. If also $\lambda > \lambda_\circ$, then the energy estimate of Theorem 2.5 gives a uniform bound of the norm of u_m in $\widetilde{W}^{1,2}(R_n^+)$. Hence some subsequence of $\{u_m\}$ converges weakly to u in $\widetilde{W}^{1,2}(R_n^+)$. By the Sobolev imbedding theorem we may assume that this subsequence converge in $L^2((K)$ to u for every compact subset $K \subset R_n^+$. It is clear that

u is a solution of (2.12). By Theorem 2.4 u has a trace $\zeta \in L^2(R_{n-1})$ on $x_n = 0$. It remains to show that $\zeta \equiv 0$ a.e. on R_{n-1}. From the proof of Theorem 2.4 we have

$$\int_{R_{n-1}} a^{nn}(x',0)\zeta(x')\Psi(x',0)\, dx = \int_{R_n^+} F(\Psi)\, dx$$

for all $\Psi \in W^{1,2}(R_n^+)$. Similarly $0 = \int_{R_n^+} F_m(\Psi)\, dx$, where F_m is obtained from F by replacing f and g^i by $T_m f$ and $T_m g^i$, respectively. Since $\lim_{m\to\infty} \int_{R_m^+} F_m(\Psi)\, dx = \int_{R_m^+} F(\Psi)\, dx$, we see that $\int_{R_m^+} a^{nn}(x',0)\zeta(x')\Psi(x',0)\, dx' = 0$ for each $\Psi \in W^{1,2}(R_n^+)$ and consequently $\zeta(x') \equiv 0$ a.e. on R_{n-1}. We now remove the restriction $\lambda > \lambda_*$ using the method of continuity. Since there is nothing to show if $\lambda_* < \lambda_\circ$, we suppose that $\lambda_* > \lambda_\circ$ and denote by B the Banach space made up of elements of $\widetilde{W}^{1,2}(R_n^+)$ for which $\lim_{\delta\to 0} \int_{R_{n-1}} u(x',\delta)^2\, dx = 0$. We shall show that there is a constant τ depending only on L such that if the problem

$$Lu + \lambda u = f - D_i g^i \text{ in } R_n^+,\ u(x',0) = 0 \text{ on } R_{n-1}$$

has a unique solution in $\widetilde{W}^{1,2}(R_n^+)$ for some $\lambda > \lambda_\circ$ and if $\sigma < \min(\tau, \lambda - \lambda_\circ)$, then the problem

$$Lu + (\lambda - \sigma)u = f - D_i g^i \text{ in } R_n^+, \quad u(x',0) = 0 \text{ on } R_{n-1}$$

is solvable in $\widetilde{W}^{1,2}(R_n^+)$. This is achieved by defining the map $T: B \to B$ by $u = Tv$ if

$$Lu + \lambda u = f + \sigma v - D_i g^i \text{ in } R_n^+, \quad u(x',0) = 0 \text{ on } R_{n-1}$$

and observing that there is a constant τ depending only on L such that T is a contraction mapping for $0 \leqq \sigma < \min(\tau, \lambda - \lambda_\circ)$. Thus T has a unique fixed point which is the solution of (2.12) provided $\lambda > \lambda_\circ$.
The penultimate step is to relax the condition on φ to $\varphi \in W^{1,2}(R_{n-1})$ (in fact, we assume that φ is a trace of an element from $W^{1,2}(R_n^+)$ and this is what we actually use). First we extend φ to a function belonging to $W^{1,2}(R_n^+)$, which we also call φ. If $v \in B$ is a solution of

$$Lv + \lambda v = f - D_i g^i - L\varphi - \lambda\varphi,$$

(note that the right–hand side of this equation lies in the appropriate space) then it is obvious that $u = v + \varphi$ is the solution of (2.12).
To obtain the theorem in its full generality, we let $\{\varphi_m\}$ be sequence of $W^{1,2}(R_{n-1})$–functions converging to φ in $L^2(R_{n-1})$. By L_m we denote the operator obtained from L by replacing b^i, d^i and c by $T_m b^i$, $T_m d^i$ and $T_m c$, respectively. By what we have already shown, the problem

$$L_m u_m + \lambda u_m = f - D_i g^i \text{ in } R_n^+, \quad u_m(x',0) = \varphi_m(x') \text{ on } R_{n-1}$$

has a solution for each m provided $\lambda > \lambda_\circ$. The energy estimate (2.13) then implies that $\{u_m\}$ has a subsequence which converges to a function u weakly in $\widetilde{W}^{1,2}(R_n^+)$ and in $L^2(K)$ for each compact subset $K \subset R_n^+$. It is obvious that u is a solution of the problem (2.12). Uniqueness of the solution is, of course, an immediate consequence of the energy estimate.

It should be noted that $\lambda_\circ$ is generally not the best constant for which Theorem 2.6 is valid. For exemple if b^i and d^i $(i = 1, ..., n)$ are in $M_\circ^n(R_n^+)$, c is in $M_\circ^{\frac{n}{2}}(R_n^+)$ and $c(x) - D_i b^i(x) \geqq \mu > 0$ in the distributional sense on R_n^+ for some constant μ, then the problem (2.12) is solvable in $\overset{\circ}{W}{}^{1,2}(R_n^+)$ (for details see [**TT**], see also [**BM**] for version of this theorem with b^i, d^i $(i = 1, ..., n)$ in $L^n(R_n^+)$ and c in $L^{\frac{n}{2}}(R_n^+)$). Using this observation we easily deduce the following existence result.

Theorem 2.7. *Suppose that b^i, d^i $(i = 1, ..., n)$ are in $M_\circ^n(R_n^+)$ and c is in $M_\circ^{\frac{n}{2}}$ and that $c(x) - D_i b^i(x) \geqq \mu > 0$ on R_n^+ (in the distributional sense), then the Dirichlet problem (2.12) (with $\lambda = 0$) has a unique solution in $\widetilde{W}^{1,2}(R_n^+)$.*

PROOF. Let us fix $\lambda \geqq \lambda_\circ$. Then the Dirichlet problem (2.12) has a unique solution $v \in \widetilde{W}^{1,2}(R_n^+)$. By the above mentioned result the Dirichlet problem

$$Lu = \lambda v \text{ in } R_n^+,\ u(x',0) = 0 \text{ on } R_{n-1}$$

has a unique solution $w \in \overset{\circ}{W}{}^{1,2}(R_n^+)$. Obviously $w + v$ is a solution of (2.12) with $\lambda = 0$. For any two solutions u_1 and u_2 in $\widetilde{W}^{1,2}(R_n^+)$ of (2.12), $z = u_1 - u_2$ is a solution of (2.12) with $z(x',0) = 0$ on R_{n-1}, $g^i \equiv 0$ $(i = 1, .., n)$ anf $f \equiv 0$. Consequently $z \in \overset{\circ}{W}{}^{1,2}(R_n^+)$ and by the result of [**TT**], $z \equiv 0$.

2.6. The Dirichlet problem in the weighted Sobolev space $W^{1,2}(R_n^+, \Psi)$.

In this section we introduce a new weighted Sobolev space $W^{1,2}(R_n^+, \Psi)$, which allows us to solve the Dirichlet problem with the boundary data having an exponential growth at infinity.

Let Ψ be a positive function in $C^1(R_{n-1})$ such that

$$\text{(W)} \qquad |D\Psi(x')| \leqq N\Psi(x') \quad \text{in} \quad R_{n-1}$$

for some constant $N > 0$.

We set

$$L^2_\Psi(R_n^+) = \{u;\quad u \in L^2_{\text{loc}}(R_n^+),\quad \int_{R_n^+} u(x)^2 \Psi(x')^2\, dx' < \infty\},$$

$$L^2_\Psi(R_{n-1}) = \{f;\quad f \in L^2_{\text{loc}}(R_{n-1}),\quad \int_{R_{n-1}} f(x')^2 \Psi(x')^2\, dx' < \infty\},$$

and

$$W^{1,2}(R_n^+; \Psi) = \{u;\quad u \in W^{1,2}_{\text{loc}}(R_n^+),\quad \int_T^\infty \int_{R_{n-1}} u(x)^2 \Psi(x')^2\, dx < \infty \quad \text{for some} \quad T > 0\}.$$

Throughout this section we suppose that (A) and (B) hold and, for simplicity, we assume that the coefficients b^i, d^i $(i = 1, ..., n)$ and c belong to $L^\infty(R_n^+)$.

First we discuss some preliminary properties of solutions in $W^{1,2}(R_n^+, \Psi)$ of the equation (2.5). We make the following assumptions on the functions f and g^i:

$$f \text{ and } g^i\,(i = 1, ..., n) \quad \text{are in} \quad L^2_\Psi(R_n^+).$$

We commence with an analogue of Lemmas 2.2 and 2.3.

Lemma 2.6. *Let $u \in W^{1,2}_\Psi(R_n^+)$ be a solution of (2.5) in R_n^+. Then for each $r > 0$ we have*

$$\int_{2r}^{\infty} |Du(x)|^2 \Psi(x')^2\, dx \leqq C\Bigg[\int_r^\infty \int_{R_{n-1}} u(x)^2 \Psi(x')^2\, dx$$
$$+ \int_r^\infty \int_{R_{n-1}} f(x)^2 \Psi(x')^2\, dx + \int_r^\infty \int_{R_{n-1}} |g(x)|^2 \Psi(x')^2\, dx\Bigg],$$

where the constant $C > 0$ depends on N, n and the bounds of the coefficients of L.

The proof is similar to that of Lemma 2.2 and therefore is omitted.

This lemma and the argument used in the proof of Theorem 2.1, with the obvious modification of the test function

$$v(x) = u(x) \max\{\mu(x_n) - \delta, 0\} \Psi(x')^2$$

yield the following result

Theorem 2.8. *Let $u \in W^{1,2}(R_n^+, \Psi)$ be a solution of (2.5) in R_n^+. Then the following conditions*

(I) $$\sup_{0<x_n<T} \int_{R_{n-1}} u(x', x_n)^2 \Psi(x')^2\, dx' < \infty \quad \text{for some } T > 0$$

and

(II) $$\int_{R_n^+} |Du(x)|^2 \Psi(x')^2 q(x_n)\, dx < \infty$$

are equivalent.

As in Section 2.4 we deduce from Theorem 2.8 the sufficient condition for the existence of traces of solutions in $W^{1,2}(R_n^+, \Psi)$.

Theorem 2.9. *Let $u \in W^{1,2}(R_n^+, \Psi)$ be a solution of (2.5) such that one of the conditions I or II of Theorem 2.8 holds. Then there exists a function $\varphi \in L^2_\Psi(R_{n-1})$ such that*

$$\lim_{s\to 0} \int_{R_{n-1}} [u(x', s) - \varphi(x')]^2 \Psi(x')^2\, dx' = 0.$$

Guided by Theorem 2.9 we adopt the following approach to the Dirichlet problem.

Let $\varphi \in L^2_\Psi(R_{n-1})$. A weak solution $u \in W^{1,2}(R_n^+, \Psi)$ of (2.5) is said to be the solution of the Dirichlet problem with the boundary condition $u(x', 0) = \varphi(x')$ if

$$\lim_{s\to 0} \int_{R_{n-1}} [u(x', s) - \varphi(x')]^2 \Psi(x')^2\, dx' = 0.$$

The existence result is based on the energy estimate.

Theorem 2.10. *There are constants d, $\lambda_\circ$ and C depending on L such that any solution of (2.12), with $\lambda > \lambda_\circ$, satisfies*

(2.15)

$$\int_{R_n^+} |Du(x)|^2 q(x_n)\Psi(x')^2\,dx + \sup_{0<s\leqq d}\int_{R_{n-1}} u(x',s)^2\Psi(x')^2\,dx'$$
$$+(\lambda-\lambda_\circ)\int_{R_n^+} u(x)^2 q(x_n)\Psi(x')^2\,dx' \leqq C\bigg[\int_{R_{n-1}} \varphi(x')^2\Psi(x')^2\,dx'$$
$$+\int_{R_n^+} f(x)^2\Psi(x')^2\,dx + \int_{R_n^+} |g(x)|^2\Psi(x')^2\,dx'\bigg].$$

The proof is identical to that of Theorem 2.5 and therefore is omitted. The energy estimate of Theorem 2.10 yields the following existence result.

Theorem 2.11. *Let C, $\lambda_\circ$ and d be the constants of Theorem 2.10 and let $\varphi \in L^2_\Psi(R_{n-1})$. If $\lambda > \lambda_\circ$, then the Dirichlet problem (2.12) has a unique solution in $W^{1,2}(R_n^+,\Psi)$.*

Obviously this solution satisfies the energy estimate of Theorem 2.10.

To show an application of Theorem 2.11, we require first the following form of the energy estimate.

Let

$$H(x,b) = \left(\prod_{i=1}^{n-1}\cosh bx_i\right)^{-1},$$

where $b > 0$, and observe that the function H satisfies (W) with N depending on b.

Theorem 2.12. *Let $\Omega \subset R_n^+$ be a bounded domain. Suppose that*

$$c(x) + \frac{1}{2}D_i b^i + \frac{1}{2}D_i d^i \geqq K \quad \text{on} \quad R_n^+ \tag{2.16}$$

for some constant $K > 0$ in the distributional sense. If $u \in \overset{\circ}{W}{}^{1,2}(\Omega)$ is the solution of the Dirichlet problem

$$Lu = f - D_i g^i \quad \text{in } \Omega \quad \text{and} \quad u(x) = 0 \quad \text{on } \partial\Omega,$$

then there exist constants b_1 and C, independent of Ω, such that

$$\int_\Omega |Du(x)|^2 H(x',b)^2\,dx + \int_\Omega u(x)^2 H(x',b)^2\,dx$$
$$\leqq C\bigg[\int_\Omega f(x)^2 H(x',b)^2\,dx + \int_\Omega |g(x)|^2 H(x',b)^2\,dx\bigg]$$

for all $0 \leqq b \leqq b_1$.

PROOF. Taking $v = uH^2$ as a test function we obtain

$$(2.17)\qquad \int_\Omega a^{ij}D_iuD_juH^2\,dx + 2\int_\Omega a^{ij}D_iuuD_jHH\,dx$$
$$+\int_\Omega (b^i+d^i)D_iuuH^2\,dx + 2\int_\Omega b^iu^2D_iHH\,dx + \int_\Omega cu^2H^2\,dx$$
$$= \int_\Omega fuH^2\,dx + \int_\Omega g^iD_iuH^2\,dx + 2\int_\Omega g^iuD_iHH\,dx.$$

Let us now observe that
(2.18)

$$\frac{1}{2}\int_\Omega (b^i+d^i)H^2D_iu^2\,dx + \int_\Omega cu^2H^2\,dx \geqq K\int_\Omega u^2H^2\,dx$$
$$-\frac{1}{2}\int_\Omega (b^i+d^i)u^2D_iH^2\,dx.$$

On the other hand by Young's inequality we have

$$(2.19)\qquad 2\int_\Omega a^{ij}D_iuuD_iHH\,dx \leqq \frac{\gamma}{3}\int_\Omega |Du|^2H^2\,dx + C_1\int_\Omega u^2|DH|^2\,dx,$$

where a constant $C_1 > 0$ depends on γ and the L^∞-bounds for a^{ij}. Similarly

$$(2.20)\qquad \int_\Omega fu^2H^2\,dx \leqq \frac{K}{2}\int_\Omega u^2H^2\,dx + \frac{2}{K}\int_\Omega f^2H^2\,dx$$

$$(2.21)\qquad \int_\Omega g^iD_iuH^2\,dx \leqq \frac{\gamma}{3}\int_\Omega |Du|^2H^2\,dx$$

$$(2.22)\qquad 2\int_\Omega g^iuD_iHH\,dx \leqq \int_\Omega |g|^2H^2\,dx + \int_\Omega u^2|DH|^2\,dx.$$

Inserting the estimates (2.18)-(2.22) into (2.17) we obtain

$$\frac{\gamma}{3}\int_\Omega |Du|^2H^2\,dx + \int_\Omega \Big[\frac{K}{2}H^2 - \frac{1}{2}(b^i+d^i)D_iH^2$$
$$+ 2b^iD_iHH - |DH|^2\Big]u^2\,dx \leqq (1+\frac{3}{\gamma})\int_\Omega |g|^2H^2\,dx + \frac{2}{K}\int_\Omega f^2H^2\,dx.$$

A direct calculation shows that there exists $b_1 > 0$ such that

$$\frac{K}{2}H^2 - \frac{1}{2}(b^i+d^i)D_iH^2 + 2b^iD_iHH - |DH|^2 \geqq \frac{K}{4}H^2$$

for all $x \in \Omega$ and $0 \leqq b \leqq b_1$ and this completes the proof.

To proceed further we set $H_1(x') = H(x', b_1)$ and denote by $\overset{\circ}{W}{}^{1,2}(R_n^+, H_1)$ the completion of the space $C_\circ^\infty(R_n^+)$ with respect to the norm

$$\|u\|_{H_1} = \int_{R_n^+} |Du(x)|^2H_1(x')^2\,dx + \int_{R_n^+} u(x)^2H_1(x')^2\,dx.$$

It is clear that

$$\int_{R_n^+}\Big[u(x)^2 + |Du(x)|^2\Big]\exp\sum_{i=1}^{n-1} b_1|x_i|\,dx < \infty$$

for all $u \in \overset{\circ}{W}{}^{1,2}(R_n^+, H_1)$.

Theorem 2.14. *Suppose that (2.16) holds and that*

$$f \in L^2_{H_1}(R_n^+) \quad \text{and} \quad g^i \in L^2_{H_1}(R_n^+),\ i = 1, ..., n.$$

Then the Dirichlet problem (2.12), with $\lambda = 0$ *and* $\varphi = 0$, *has a unique solution in* $\overset{\circ}{W}{}^{1,2}(R_n^+, H_1)$.

PROOF. Let $\{\Omega_m\}$ be an increasing sequence of bounded domains, with smooth boundaries $\partial\Omega_m$, such that $\bigcup_{m \geqq 1} \Omega_m = R_n^+$. For each $m \geqq 1$ the Dirichlet problem

$$Lu = f - D_i g^i \quad \text{in } \Omega_m, \quad u(x) = 0 \quad \text{on } \partial\Omega_m$$

has a unique solution $u_m \in \overset{\circ}{W}{}^{1,2}(\Omega_m)$. We now extend $u_m(x)$ by 0 for $x \notin \Omega_m$. The resulting sequence is in $\overset{\circ}{W}{}^{1,2}(R_n^+)$ and satisfies the estimate of Theorem 2.12. We complete the proof by selecting a subsequence converging to a function u weakly in $\overset{\circ}{W}{}^{1,2}(R_m^+, H_1)$ and strongly in $L^2(K)$ for each compact set $K \subset R_n^+$.

Theorem 2.14. *Let* $\varphi \in L^2_{H_1}(R_{n-1})$, $f \in L^2_{H_1}(R_n^+)$ *and* $g^i \in L^2_{H_1}(R_n^+)$, $i = 1, ..., n$. *Suppose that the coefficients of* L *satisfy (2.15). Then the Dirichlet problem (2.12), with* $\lambda = 0$, *has a unique solution in* $W^{1,2}(R_n^+, H_1)$.

The proof is similar to that of Theorem 2.7 and is based on Theorems 2.11 and 2.13.

CHAPTER 3

THE DIRICHLET PROBLEM IN A BOUNDED DOMAIN

The purpose of this chapter is to extend the method described in Chapter 2 to a bounded domain with C^2-boundary. This is achieved by flattening locally the boundary and approximating the coefficients a^{ij}, satisfying a Dini condition, by coefficients having partial derivatives. This chapter also contains an investigation of traces of $\widetilde{W}^{12}$-solutions. This leads to a proper formulation of the boundary condition of the Dirichlet problem with L^2-boundary data. Again, as in Chapter 2, the main point in our approach to the Dirichlet problem is the energy estimate.

3.1. Assumptions and preliminaries.

Throughout this chapter we assume that Q is a bounded domain with C^2-boundary. In Q we consider the equation

$$Lu + \lambda u = f, \tag{3.1}$$

where L is the elliptic operator given by (1.1) and λ is a real parameter. We make the following assumptions on the coefficients of the operator L.

(A_1) There exists a positive constant γ such that

$$a^{ij}(x)\xi_i\xi_j \geqq \gamma|\xi|^2$$

for all $\xi \in R_n$ and $x \in Q$, where $a^{ij} \in L^\infty(Q)\,(i,j = 1,...,n)$, $a^{ij} = a^{ji}$, $(i,j = 1,...,n)$.

(A_2) There is a non-decreasing function ω on $[0,\infty)$ such that

$$|a^{ij}(x) - a^{ij}(y)| \leqq \omega(|x-y|)^2 \; (i,j = 1,...,n)$$

for all $x, y \in Q$, and

$$\int_0^1 \omega(t)t^{-1}\,dt < \infty.$$

(A_3) There are τ, τ_1, τ_2, β and $\beta_1 \in (0,1)$ and $\beta_2 \in (0,2)$ such that

$$b^i + d^i = e_1^i + e_2^i \text{ with } e_1^i r^{-\tau} \in L^n(Q),\; e_2^i r^{\beta} \in L^\infty(Q),$$
$$b^i = b_1^i + b_2^i \text{ with } b_1^i r^{-\tau_1} \in L^n(Q),\; b_2^i r^{\beta_1} \in L^\infty(Q), \text{ and}$$
$$c = c_1 + c_2 \text{ with } c_1 r^{-\tau_2} \in L^{\frac{n}{2}}(Q),\; c_2 r^{\beta_2} \in L^\infty(Q),$$

where $r(x) = \operatorname{dist}(x, \partial Q)$.

(A_4) There is a constant θ with $2 \leqq \theta < 3$ such that

$$\int_Q f(x)^2 r(x)^\theta\,dx < \infty.$$

Lemma 3.1 below, is an obvious version of Lemma 2.1.

Lemma 3.1. *If $gr^{-\mu} \in L^s(Q)$, $1 < s < \infty$, $\mu > 0$ and $\epsilon > 0$, then g admits a decomposition $g = g_1 + g_2$ such that*

$$\|g_1 r^{-\mu}\|_s \leqq \epsilon \quad \text{and} \quad g_2 r^{-\mu} \in L^\infty(Q).$$

Lemma 3.2. *Let $0 < \delta_2 < \delta_1 < \delta_\circ$. If $u \in W^{1,2}_{\text{loc}}(Q)$ is a solution of (3.1), then*

$$\int_{Q_{\delta_1}} |Du(x)|^2\, dx + \lambda \int_{Q_{\delta_1}} |u(x)|^2\, dx \leqq M\left[\int_{Q_{\delta_2}} f(x)^2\, dx + \int_{Q_{\delta_2}} u(x)^2\, dx\right],$$

where a positive constant M depends on the norm of the coefficients of L in Q_{δ_2}, γ, n, δ_1 and δ_2.

PROOF. Let $\Phi \in C^1(Q)$ with supp $\Phi \in Q_{\delta_2}$, $\epsilon > 0$ and set

$$v(x) = u(x)\Phi(x)^2.$$

Substituting v into (3.1), we obtain

$$\begin{aligned}(3.2) \qquad & \int_{Q_{\delta_2}} a^{ij} D_i u D_j u \Phi^2\, dx + 2\int_{Q_{\delta_2}} a^{ij} D_i u u \Phi D_j \Phi\, dx \\ & + \int_{Q_{\delta_2}} (b^i + d^i) u D_i u \Phi^2\, dx + 2\int_{Q_{\delta_2}} b^i u^2 \Phi D_j \Phi\, dx + \int_{Q_{\delta_2}} c u^2 \Phi^2\, dx \\ & + \lambda \int_{Q_{\delta_2}} u^2 \Phi^2\, dx - \int_{Q_{\delta_2}} f u \Phi^2\, dx = 0.\end{aligned}$$

Let us denote the integrals in (3.2) by $j_1, \ldots, j_7$. In the estimates, below, $C_1, C_2, \ldots$ are positive constants, which depend on the norms of the coefficients in Q_{δ_2}, γ and n. By (A_1) and the Young inequality we obtain

$$j_1 \geqq \int_{Q_{\delta_2}} |Du|^2 \Phi^2\, dx$$

and

$$|j_2| \leqq \frac{\gamma}{7} \int_{Q_{\delta_2}} |Du|^2 \Phi^2\, dx + C_1 \int_{Q_{\delta_2}} u^2 |D\Phi|^2\, dx.$$

Using Lemma 3.1 and (A_3), we obtain

$$j_3 = \int_{Q_{\delta_2}} g_1^i D_i u u \Phi^2\, dx + \int_{Q_{\delta_2}} (g_2^i + e_2^i) D_i u u \Phi^2\, dx = j_3' + j_3'',$$

with $\|g_1^i\|_{n,Q_{\delta_2}} \leqq \epsilon_1$ and $g_2^i \in L^\infty(Q_{\delta_2})$, where $\epsilon_1 = \frac{2\gamma}{21S}$ and S is a constant from the Sobolev inequality. Hölder's, Sobolev's and Young's inequalities yield

$$\begin{aligned}|j_3'| &\leqq \|g_1^i\|_{n,Q_{\delta_2}} \|Du\Phi\|_{2,Q_{\delta_2}} S \|D(u\Phi)\|_{2,Q_{\delta_2}} \\ &\leqq \frac{\gamma}{7} \int_{Q_{\delta_2}} |Du|^2 \Phi^2\, dx + C_2 \int_{Q_{\delta_2}} u^2 |D\Phi|^2\, dx,\end{aligned}$$

and

$$|j_3''| \leqq \frac{\gamma}{7} \int_{Q_{\delta_2}} |Du|^2 \Phi^2 \, dx + C_3 \int_{Q_{\delta_2}} u^2 \Phi^2 \, dx.$$

Similarly we get for $k = 4, 5, 6$

$$|j_k| \leqq \frac{\gamma}{7} \int_{Q_{\delta_2}} |Du|^2 \Phi^2 \, dx + C_k \int_{Q_{\delta_2}} u^2 \left(|D\Phi|^2 + \Phi^2 \right) dx,$$

and

$$|j_7| \leqq \frac{1}{2} \int_{Q_{\delta_2}} f^2 \Phi^2 \, dx + \frac{1}{2} \int_{Q_{\delta_2}} u^2 \Phi^2 \, dx.$$

Inserting these estimates into (3.2) we obtain

$$\int_{Q_{\delta_1}} |Du|^2 \Phi^2 \, dx + \lambda \int_{Q_{\delta_1}} u^2 \Phi^2 \, dx$$
$$\leqq C_7 \left[\int_{Q_{\delta_2}} u^2 \left(\Phi^2 + |D\Phi|^2 \right) dx + \int_{Q_{\delta_2}} f^2 \Phi^2 \, dx \right].$$

Now take $\Phi = \Phi_m$, where Φ_m is an increasing sequence of non-negative functions in $C^1(Q)$, converging to 1 on Q_{δ_1} as $m \to \infty$, vanishing outside Q_{δ_2} and with the gradient bounded independently of m. Then let $m \to \infty$ and the result easily follows.

To proceed further we define a function $s_\delta : Q - Q_{\delta_o} \to \partial Q_\delta$ by

$$s_\delta(x) = x + (r(x) - \delta) \nu_{r(x)}(x),$$

where $\nu_{r(x)}$ denotes the outer normal to $Q_{r(x)}$ at x and $\delta \in [0, \delta_o)$. Moreover, let p_n be the projection from R_n onto R with respect to the n-th component, that is, $p_n(x) = x_n$ for $(x', x_n) \in R_{n-1} \times R$.

Lemma 3.3. *There exists a positive constant $\delta_* < \delta_o$, a finite open covering $\{U_1, ..., U_N\}$ of $\overline{Q - Q_{\delta_*}}$ in R_n, and functions $\theta_{\delta,k}$ such that*

(1) *$\theta_{\delta,k}$ are differentiable functions from $\overline{U_k} \cap \overline{Q_\delta}$ to R_n^+ with uniformly bounded gradients for all $\delta \in [0, \delta_*)$,*

(2) *$\theta_{\delta,k}$ are one - to - one; the inverse functions are Lipschitz continuous functions on the image of $\theta_{\delta,k}$ with a uniform Lipschitz constant for all $\delta \in [0, \delta_*)$,*

(3) *(image $\theta_{\delta,k}$)$\cap(R^{n-1} \times \{0\}) = \theta_{\delta,k}(U_k \cap \partial Q_\delta)$,*

(4) *$|\theta_{\delta,k} - \theta_{\delta,k} \circ s_\delta| = p_n \circ \theta_{\delta,k} = r(.) - \delta$ in $U_k \cap Q_\delta$ for all $\delta \in [0, \delta_*)$ and $k = 1, ..., N$.*

PROOF. Choosing $\delta_* \in (0, \delta_o)$ sufficiently small and suitable covering $\{U_1, ..., U_N\}$ of open sets in R_n with smooth boundaries, we may assume that there are hyperplanes H_k in R_n, which are disjoint with $\bar{Q}$, and that $\partial Q \cap U_k$ are given by C^2- functions σ_k, which are defined on H_k, $k = 1, ..., N$. Using rotations and translations, if necessary, we can assume, without loss of generality, that $H_k = R_{n-1}$ and $Q \subset R_n^+$. We define

$$\theta_{\delta,k}(x) = \Big((s_\circ(x))', r(x) - \delta\Big),$$

for each $x \in U_k \cap \bar{Q}_\delta$ and $\delta \in [0, \delta_*)$. It is easy to see that

$$\theta_{\delta,k}^{-1}(y) = (y', \sigma_k(y')) - (y_n + \delta)\nu(y', \sigma_k(y'))$$

for all y in the image of $\theta_{\delta,k}$. Here $\nu(y', \sigma_k(y'))$ denotes the outward normal to ∂Q at $(y', \sigma_k(y'))$. Hence (1), (2) and (3) hold. Finally, we observe that

$$\begin{aligned} |\theta_{\delta,k}(x) - \theta_{\delta,k}(s_\delta(x))| &= |((s_\circ(x))', r(x) - \delta) - ((s_\circ(x)', 0)| \\ &= |r(x) - \delta| = p_n(\theta_{\delta,k}(x)) \end{aligned}$$

for each $x \in U_k \cap Q_\delta$ and this completes the proof.

Lemma 3.4. *There are functions $A_\delta^{ij} \in C^1(Q_\delta)$, positive constants K_1, K_2 and non-decreasing function $\tilde{\omega}$ defined on $[0, \infty)$, such that*

(1) $|A_\delta^{ij}(x) - a^{ij}(x)| \leqq \tilde{\omega}(r(x) - \delta)^2$,

(2) $|D_i A_\delta^{ij}(x)| \leqq K_1 \tilde{\omega}(r(x) - \delta)^2 (r(x) - \delta)^{-1}$.

(3) $\int_0^1 \tilde{\omega}(t) t^{-1}\, dt < \infty$,

(4) $A_\delta^{ij}(x)\xi_i\xi_j \geqq \gamma|\xi|^2$, $\|A_\delta^{ij}\|_\infty \leqq M_\circ$ *and* $A_\delta^{ij} = A_\delta^{ji}$ *for all* $x \in Q_\delta$, $\delta \in [0, \delta_*)$, $\xi \in R_n$ *and* $i, j = 1, ..., n$, *where* δ_* *is given in Lemma 3.3 and* $M_\circ = \max\{\|a^{ij}\|_\infty ij = 1, ..., n\}$,

(5) *Moreover, for all* $\epsilon > 0$ *there is a* $\hat{\delta} \in (0, \delta_*)$ *such that* $|A_\delta^{ij}(x) - A_\circ^{ij}(x)| < \epsilon$ *for all* $\delta \in (0, \hat{\delta})$, $x \in Q_\delta$ *and* $i, j = 1, ..., n$.

PROOF. First we construct Dini extensions of the functions a^{ij} into R_n. Let $\{V_1, ..., V_M\}$ be an open and convex covering of ∂Q such that $V_k \cap Q$ is star-shaped with respect to some center point z_k in $V_k \cap Q$ $(k = 1, ..., M)$. For $y \in V_k - Q$ let $\Delta_{y,k}$ be the straight line through y and z_k, and $\pi_k(y)$ be the point of intersection of this line with ∂Q with shortest distance to y. Choosing a finer covering, if necessary, we can assume that angle between the line $\Delta_{y,k}$ and the tangent hyperplane $T_{\pi_k}(y)$ to ∂Q at $\pi_k(y)$ is larger than some fixed positive constant $\epsilon_\circ$, uniformly for all $y \in V_k - Q$ and $k = 1, ..., M$. For $y \in V_k \cap Q$ we set $\pi_k(y) = y$. Since a^{ij} can be extended continuously into $\bar{Q}$, we may assume that a^{ij} are defined on $\bar{Q}$. Obviously the Dini condition (A_2) remains valid on $\bar{Q}$. We now set

$$\hat{a}^{ij}(x) = \begin{cases} a^{ij}(x) & \text{for } x \in Q, \\ \sum_{k=1}^M a^{ij}(\pi_k(x)) g_k(x) & \text{elsewhere}, \end{cases}$$

where $(g_1, ..., g_M)$ is a C^1-partition of unity subordinate to $\{V_1, ..., V_M\}$ and π_k is defined to be arbitrary outside V_k. Moreover, we set

$$\omega_1(t) = C_1 t^{\frac{1}{2}} + \omega(C_2 t),$$

where $C_1 > 0$ and $C_2 \geqq 1$ are constants, which will be determined later. It is clear that ω_1 satisfies (3). We now claim that

$$|\hat{a}^{ij}(x) - \hat{a}^{ij}(y)| \leqq \omega_1(|x-y|)^2 \tag{3.3}$$

for all $x, y \in R_n^+$. Obviously this inequality holds true on Q because $\hat{a}^{ij} = a^{ij}$ on Q. Therefore, we first consider the case $x, y \in R_n^+ - Q$. Observe that π_k $(k = 1, ..., M)$ are Lipschitz continuous functions. Thus letting C_2 be larger than the Lipschitz constant of π_k, and C_1^2 be larger than $M_\circ \max\{\|Dg_k\|_\infty k = 1, ..., M\}$ we obtain by the mean value theorem

$$|\hat{a}^{ij}(x) - \hat{a}^{ij}(y)| = |\sum_{k=1}^{M} a^{ij}(\pi_k(x))(g_k(x) - g_k(y))$$

$$+ \sum_{k=1}^{M}((a^{ij}(\pi_k(x)) - a^{ij}(\pi_k(y)))g_k(y)| \leqq C_1^2|x-y| + \omega(C_2|x-y|)^2$$

$$\leqq \omega_1(|x-y|)^2.$$

Finally, let us consider the case $x \in Q$ and $y \in V_k - Q$. Since the angle between the line $\Delta_{y,k}$ and the hyperplane $T_{\pi_k(y)}$ is larger than $\epsilon_\circ$ it is easy to see that

$$|\pi_k(y) - x||y - x|^{-1} \leqq C_3,$$

where C_3 is some positive constant, which is independent of x and y. Thus letting C_2 to be larger than C_3, we obtain

$$|\hat{a}^{ij}(x) - \hat{a}^{ij}(y)| \leqq \sum_{k=1}^{M} |a^{ij}(x) - a^{ij}(\pi_k(y)||g_k(y)|$$

$$\leqq \sum_{k=1}^{M} \omega(|x - \pi_k(y)|)^2 g_k(y) \leqq \omega(C_3|x-y|)^2 \leqq \omega_1(|x-y|)^2.$$

On the other hand, there is a positive constant ϵ_1 such that

$$\text{dist}\,(y, \partial Q) > \epsilon_1$$

for all $y \in R_n - \left(Q \cup \bigcup_{k=1}^M V_k\right)$. Letting $C_1 \geqq M_\circ \epsilon_1^{-1}$, we obtain

$$|\hat{a}^{ij}(x) - \hat{a}^{ij}(y)| = |a^{ij}(x)| \leqq \|a^{ij}\|_\infty \epsilon^{-1}|x-y| \leqq \omega_1(|x-y|)^2$$

for all $x \in Q$ and $y \in R_n^+ - \left(Q \cup \bigcup_{k=1}^M V_k\right)$. This completes the proof of the inequality (3.3).

Let $\{U_1, ..., U_N\}$ and $\theta_{\delta,k}$ be as in Lemma 3.3. We extend $\theta_{\delta,k}^{-1}$ to a Lipschitz continuous function $\tilde{\theta}_{\delta,k}$ on R_n^+ with a uniform constant C_4, that is,

$$|\tilde{\theta}_{\delta,k}(z) - \tilde{\theta}_{\delta,k}(\bar{z})| \leqq C_4|z - \bar{z}| \tag{3.4}$$

for all $z, \bar{z} \in R_n^+$, $\delta \in [0, \delta_*)$ and $k = 1, ..., N$. Moreover, let $\{f_1, ..., f_N\}$ be a C^1-partition of unity subordinate to $\{U_1, ..., U_N\}$ and g be a C^∞-function from R into R with support in $[-1, 1]$ such that $\int_{-1}^1 g(t)\, dt = 1$. For $t \geqq 0$, $z \in R_n^+$ and $x \in Q_\delta$ we define

$$\begin{aligned}
&\omega_2(t) = \omega_1(C_4 t),\\
&\tilde{\omega}(t) = \omega_2(t) + t^{\frac{1}{2}},\\
&a^{ij}_{\delta,k}(z) = \hat{a}^{ij}(\tilde{\theta}_{\delta,k}(z)),\\
&A^{ij}_{\delta,k}(z) = \int_{-1}^1 a^{ij}_{\delta,k}(z_1, ..., z_{i-1}, z_i - tz_n, z_{i+1}, ..., z_n) g(t)\, dt,\\
&A^{ij}_\delta(x) = \sum_{k=1}^N A^{ij}_{\delta,k}(\theta_{\delta,k}(x))\, f_k(x),
\end{aligned}$$

where $\theta_{\delta,k}$ may be defined arbitrarily in $Q_\delta - U_k$. Observe that $\tilde{\omega}$ satisfies (3) and that (4) holds. By virtue of (3.3) and (3.4) we obtain

$$|a^{ij}_{\delta,k}(z) - a^{ij}_{\delta,k}(\bar{z})| \leqq \omega_1\Big(|\tilde{\theta}_{\delta,k}(z) - \tilde{\theta}_{\delta,k}(\bar{z})|\Big)^2 \leqq \omega_2(|z - \bar{z}|)^2$$

for all $\delta \in [0, \delta_*)$. Consequently, we have

$$\begin{aligned}
|a^{ij}_{\delta,k}(z) - A^{ij}_{\delta,k}(z)| &\leqq \int_{-1}^1 |a^{ij}_{\delta,k}(z) - a^{ij}_{\delta,k}(z_1, ..., z_{i-1}, z_i - tz_n, z_{i+1}, ..., z_n)| g(t)\, dt\\
&\leqq \int_{-1}^1 \omega_2(|z_n t|)^2 g(t)\, dt \leqq \omega_2(|z_n|)^2
\end{aligned} \tag{3.5}$$

for $\delta \in [0, \delta_*)$ and $k = 1, ..., N$. It follows from (3.5) and the relation (4) of Lemma 3.3 that

$$\begin{aligned}
|A^{ij}_\delta(x) - a^{ij}(x)| &\leqq \sum_{k=1}^N |A^{ij}_{\delta,k}(\theta_{\delta,k}(x)) - a^{ij}_{\delta,k}(\theta_{\delta,k}(x))| f_k(x)\\
&\leqq \omega_2(|p_n(\theta(x))|)^2 \leqq \tilde{\omega}(r(x) - \delta)^2
\end{aligned}$$

for $x \in Q_\delta$, $\delta \in [0, \delta_*)$ and $k = 1, .., N$ and this proves the relation (1). To show (2), we first observe that $\int_{-1}^1 g'(t)\, dt = 0$ and consequently

$$|D_i A^{ij}_{\delta,k}(z)| \leqq C_5 \omega_2(z_n)^2 z_n^{-1} \tag{3.6}$$

for all $z \in R_n^+$ and $i, j = 1, ..., n$, where $C_5 = \int_{-1}^{1} |g'(t)|\, dt$. From (3.6) and the relation (4) of Lemma 3.3 we deduce

$$|D_i A_\delta^{ij}(x)|$$
$$\leqq K_1 \Big(\omega_2(r(x) - \delta)^2 (r(x) - \delta)^{-1} + 1 \Big) \leqq K_1 \tilde{\omega}(r(x) - \delta)^2 (r(x) - \delta)^{-1},$$

with K_1 depending on C_5, $M_\circ$, N, $|||Df|||_\infty$ and $|||D\theta_{\delta,k}|||_\infty$.
To show (5), we notice that from the definition of $\theta_{\delta,k}$ it follows that

$$|p_i(\theta_{\delta,k}(x) - \theta_{0,k}(x))| = \begin{cases} 0 & \text{for } i \neq n \\ \delta & \text{for } i = n, \end{cases}$$

for all $x \in \bigcup_{k=1}^N \Big(Q_\delta \cap U_k \Big)$. Thus, by virtue of (3.3) and (3.4) we find

$$|A_\delta^{ij}(x) - A_\circ^{ij}(x)|$$
$$= |\sum_{k=1}^N \int_{-1}^1 \Big[\hat{a}^{ij} \Big(\tilde{\theta}_{\delta,k}((p_1\theta_{\delta,k}(x), ..., p_i\theta_{\delta,k}(x) - tp_n\theta_{\delta,k}(x), ..., p_n\theta_{\delta,k}(x)) \Big)$$
$$- \hat{a}^{ij} \Big(\tilde{\theta}_{0,k}((p_1\theta_{0,k}(x), ..., p_i\theta_{0,k}(x) - tp_n\theta_{0,k}(x), ...p_n\theta_{0,k}(x))) \Big) f_k(x) \Big] g(t)\, dt|$$
$$\leqq \sum_{k=1}^N \int_{-1}^1 \Big[\omega_2 \Big(|(p_1(\theta_{\delta,k}(x) - \theta_{0,k}(x)), ..., p_i(\theta_{\delta,k}(x) - \theta_{0,k}(x)) - tp_n(\theta_{\delta,k}(x)$$
$$- \theta_{0,k}(x)), ..., p_n(\theta_{\delta,k}(x) - \theta_{0,k}(x))| \Big) \Big] g(t)\, dt f_k(x) \leqq \int_{-1}^1 \omega_2(\delta(t^2+1)^{\frac{1}{2}}) g(t)\, dt$$
$$\leqq \omega_2(\delta 2^{\frac{1}{2}}),$$

for $x \in Q_\delta$, which proves the relation (5).

From now on we assume that $\delta_* = \delta_\circ$ and abbreviate $A_\circ^{ij} = A^{ij}$. We also require the following obvious result

Lemma 3.5. *If $u \in L^2_{\text{loc}}(Q)$ is such that*

$$\sup_{0<\delta<\delta_\circ} \int_{\partial Q} u(x)^2\, dS_x < \infty,$$

then for each $0 < d \leqq \delta_\circ$ and $0 \leqq \mu < 1$, we have

$$\int_{Q-Q_\delta} u(x)^2 r(x)^{-\mu}\, dx \leqq \frac{d^{1-\mu}}{1-\mu} \sup_{0<\delta<d} \int_{\partial Q_\delta} u(x)^2\, dS_x, \tag{3.7}$$

and

$$\int_{Q-Q_\delta} u(x)^2 \tilde{\omega}(r(x))^2 r(x)^{-1}\, dx \leqq \tilde{\omega}(d) \int_0^d \frac{\tilde{\omega}(t)}{t}\, dt \sup_{0<\delta<d} \int_{\partial Q_\delta} u(x)^2\, dS_x. \tag{3.8}$$

3.2. Weak convergence of solutions at the boundary.

In this section we prove some preliminary results that will be used in the next section to establish a sufficient condition for the existence of traces of solutions in $W^{1,2}_{loc}(Q)$.

Theorem 3.1. *If $u \in W^{1,2}_{\mathrm{loc}}(Q)$ is a solution of (1.1) then the following conditions*

(I)
$$\sup_{0<\delta<\delta_\circ} \int_{\partial Q_\delta} u(x)^2 \, dS_x < \infty,$$

and

(II)
$$\int_Q |Du(x)|^2 r(x)\, dx < \infty$$

are equivalent.

PROOF. We start with the proof of $(I) \Rightarrow (II)$. Let $\delta_1 \in (0, \frac{1}{2}\delta_\circ)$, $\delta \in (0, \delta_1)$, $\epsilon > 0$ and Φ be a smooth function on $\bar{Q}$ with properties: $\Phi(x) = 1$ in $Q - Q_{\frac{1}{2}\delta_\circ}$, $\Phi(x) = 0$ in Q_{δ_1} and $0 \leqq \Phi \leqq 1$ on Q. We now define

$$\rho_1(x) = \begin{cases} 0 & \text{for } x \in Q - Q_\delta, \\ (\rho(x)^{\frac{1}{2}} - \delta^{\frac{1}{2}})^2 & \text{for } x \in Q_\delta - Q_{\delta_\circ}, \\ (\delta_\circ^{\frac{1}{2}} - \delta^{\frac{1}{2}})^2 & \text{for } x \in Q_{\delta_\circ}, \end{cases}$$

where ρ is a function described in Section 1.4 (Chapter 1), and

$$v_1(x) = \begin{cases} u(x)\rho_1(x)\Phi(x)^2 & \text{for } x \in Q_\delta, \\ 0 & \text{elsewhere}. \end{cases}$$

Then v_1 is a legitimate test function in (1.1) and on substitution we obtain

(3.9)
$$\begin{aligned} &\int_{Q_\delta} a^{ij} D_i u D_j u \rho_1 \Phi^2\, dx + \int_{Q_\delta} [a^{ij} - A^{ij}] D_j u u D_i \rho_1 \Phi^2\, dx \\ &+ \frac{1}{2}\int_{Q_\delta} A^{ij} D_j u^2 D_i \rho_1 \Phi^2\, dx + 2\int_{Q_\delta} a^{ij} D_j u u \rho_1 \Phi D_i \Phi\, dx \\ &+ \int_{Q_\delta} (d^i + b^i) D_i u u \rho_1 \Phi^2\, dx + \int_{Q_\delta} b^i u^2 D_i \rho_1 \Phi^2\, dx + 2\int_{Q_\delta} b^i u^2 \rho_1 \Phi D_i \Phi\, dx \\ &+ \int_{Q_\delta} c u^2 \rho_1 \Phi^2\, dx - \int_{Q_\delta} f u \rho_1 \Phi^2\, dx = 0. \end{aligned}$$

Let us denote the integrals on the left hand-side of (3.9) by $J_1, ..., J_9$. In the sequel positive constants which are independent of δ and u are denoted by $C_1, C_2, ...$ Moreover, we set $M_1 = \max\{\|D_i\rho\|_{\infty, Q-Q_{\delta_1}};\ \ i = 1, ..., n\}$. It follows from (A_1) that

$$J_1 \geqq \gamma \int_{Q_\delta} |Du|^2 \rho_1 \Phi^2\, dx.$$

Since $|D\rho_1|^2\rho = \rho_1|D\rho|^2$ in $Q_\delta - Q_{\delta_o}$, the Young inequality and the inequality (1) from Lemma 3.4 imply that

$$(3.10) \qquad |J_2| \leqq D_1 \int_{Q_\delta} u^2(\tilde{\omega} \circ r)^2 r^{-1}\Phi^2\,dx + \frac{\gamma}{7}\int_{Q_\delta} |Du|^2\rho_1\Phi^2\,dx,$$

with C_1 depending on γ, $\tilde{\omega}(\delta_1)$ and M_1. We now note that

$$\int_{Q_\delta - Q_{\delta_o}} \sum_{i,j=1}^{n} |D_iD_j\rho_1|\,dx$$

is uniformly bounded for all $\delta \in (0, \delta_1)$ by some $M_2 > 0$. Therefore integrating by parts and using the relation (2) from Lemma 3.4 we obtain

$$(3.11) \quad |J_3| \leqq C_3\Bigg[\sup_{0<s<\delta_o} \int_{\partial Q_\delta} u(x', s)^2\,dS_x$$
$$+ \int_{Q_\delta} u^2\Big(\Phi^2 + (\tilde{\omega} \circ r)^2 r^{-1}\Phi^2 + |D\Phi|^2\Big)\,dx\Bigg].$$

Here C_2 depends on M_o, M_1 and M_2. Applying Young's inequality we get

$$(3.12) \qquad |J_4| \leqq \frac{\gamma}{8}\int_{Q_\delta} |Du|^2\rho_1\Phi^2\,dx + C_3\int_{Q_\delta} u^2|D\Phi|^2\,dx,$$

where C_3 depends on M_o and γ. By Lemma 3.1 and (A_3) we obtain

$$J_5 = \int_{Q_\delta} g_1^iD_iuu\rho_1\Phi^2\,dx + \int_{Q_\delta} (g_2^i + e_2^i)D_iuu\rho_1\Phi^2\,dx = J_5' + J_5'',$$

where

$$e_1^i = g_1^i + g_2^i, \quad \sum_{i=1}^{n} \|g_1^ir^{-\tau}\|_{n,Q} \leqq \epsilon_1, \quad g_2^ir^{-\tau} \in L^\infty(Q),$$

with $\epsilon_1 > 0$ to be determined later. By virtue of Hölder's and Sobolev's inequality we have

$$|J_5'| \leqq \epsilon_1 S\|Du\rho_1^{\frac{1}{2}}\Phi\|_{2,Q_\delta}\Big[\|Du\rho_1^{\frac{1}{2}}\Phi r^\tau\|_{2,Q_\delta}$$
$$+ \|uD(\rho_1^{\frac{1}{2}}r^\tau)\Phi\|_{2,Q_\delta} + \|u\rho_1^{\frac{1}{2}}r^\tau|D\Phi|\|_{2,Q_\delta}\Big].$$

Since

$$D_i(\rho_1^{\frac{1}{2}}r^\tau) = \Big((\tau + \frac{1}{2})\rho^{\tau-\frac{1}{2}} - \delta^{\frac{1}{2}}\tau\rho^{\tau-1}\Big)D_j\rho$$

in $Q_\delta - Q_{\delta_o}$, we obtain

$$|D_j(\rho_1^{\frac{1}{2}}r^\tau)|^2 \leqq C_4\rho^{2\tau-1},$$

where C_4 depends on τ and M_1. Consequently, choosing ϵ_1 sufficiently small and applying Young's inequality we get

$$(3.13) \quad |J_5'| \leqq \frac{\gamma}{8} \int_{Q_\delta} |Du|^2 \rho_1 \Phi^2\, dx + C_5 \left[\int_{Q_\delta} u^2 r^{2\tau-1} \Phi^2\, dx + \int_{Q_\delta} u^2 r^{1+2\tau} |D\Phi|^2\, dx \right],$$

and

$$(3.14) \qquad |J_5''| \leqq \frac{\gamma}{8} \int_{Q_\delta} |Du|^2 \rho_1 \Phi^2\, dx + C_6 \int_{Q_\delta} u^2 (r^{1+2\tau} + r^{1-2\beta}) \Phi^2\, dx,$$

where C_5 and C_6 depend on τ, M_1, n, γ, δ_1, $\|g_2^i r^{-\tau}\|_{\infty, Q-Q_{\delta_1}}$ and $\|e_2^i r^\beta\|_{\infty, Q-Q_{\delta_1}}$ $(i = 1, ..., n)$. To estimate J_6 we again use Lemma 3.1 to get a decomposition of $b_1^i = h_1^i + h_2^i \quad (i = 1, ..., n)$ with

$$\sum_{i=1}^n \|h_1^i r^{-\tau_1}\|_{n,Q} \leqq \epsilon_2, \quad h_2^i r^{-\tau_1} \in L^\infty(Q)$$

and $\epsilon_2 > 0$. Consequently we obtain

$$\begin{aligned} |J_6| &\leqq \int_{Q_\delta} |h_1^2 u^2 D_i \rho_1 \Phi^2|\, dx + \int_{Q_\delta} |(h_2^i + b_2^i) u^2 D_i \rho_1| \Phi^2\, dx \\ &= J_6' + J_6''. \end{aligned}$$

The estimate for J_6' follows from the Sobolev and Hölder inequalities

$$(3.15) \quad \begin{aligned} J_6' &\leqq S\epsilon_2 \|u r^{\frac{\tau_1}{2}} |D\rho_1| \rho_1^{-\frac{1}{2}} \Phi\|_{2,Q_\delta} \|D(u r^{\frac{\tau_1}{2}} \rho_1^{\frac{1}{2}} \Phi)\|_{2,Q_\delta} \\ &\leqq \frac{\gamma}{8} \int_{Q_\delta} |Du|^2 \rho_1 \Phi^2\, dx + C_7 \int_{Q_\delta} u^2 (r^{\tau_1 - 1} \Phi^2 + r^{1+\tau_1} |D\Phi|^2)\, dx. \end{aligned}$$

Also, we have

$$J_6'' \leqq C_8 \int_{Q_\delta} u^2 (r^{\tau_1} + r^{-\beta_1}) \Phi^2\, dx.$$

Similarly

$$|J_7| \leqq \frac{\gamma}{8} \int_{Q_\delta} |Du|^2 \rho_1 \Phi^2\, dx + C_9 \int_{Q_\delta} u^2 (r^{\tau_1 - 1} + r^{1+\tau_1} + r^{1-\beta_1})(\Phi^2 + |D\Phi|^2)\, dx.$$

The constants C_7, C_8 and C_9 depend on τ_1, M_1, n, γ_1, $\|h_2^i r^{-\tau_1}\|_{\infty, Q-Q_\delta}$ and $\|b_2^i r^{\beta_1}\|_{\infty, Q-Q_{\delta_1}}$. We now write

$$J_8 = \int_{Q_\delta} \bar{c} u^2 \rho_1 \Phi^2\, dx + \int_{Q_\delta} (\tilde{c} + c_2) u^2 \rho_1 \Phi^2\, dx = J_8' + J_8'',$$

where

$$\|\bar{c}r^{-\tau_2}\|_{\frac{n}{2}} \leqq \epsilon_3 \quad \text{and} \quad \tilde{c}r^{-\tau_2} \in L^\infty(Q).$$

Then we have

(3.16)

$$\begin{aligned}|J_8'| &\leqq \epsilon_3 S^2 \|D(u\rho_1^{\frac{1}{2}} r^{\frac{\tau_2}{2}} \Phi)\|_{2,Q_\delta}^2 \\ &\leqq \frac{\gamma}{8} \int_{Q_\delta} |Du|^2 \rho_1 \Phi^2 \, dx + C_{10} \int_{Q_\delta} u^2 (r^{\tau_2 - 1} \Phi^2 + r^{1+\tau_2} |D\Phi|^2) \, dx\end{aligned}$$

for suitably chosen $\epsilon_3 > 0$, and

$$|J_8''| \leqq C_{11} \int_{Q_\delta} u^2 (r^{1+\tau_2} + r^{1-\beta_2}) \Phi^2 \, dx. \tag{3.17}$$

The constants C_{10} and C_{11} depend on τ_2, M_1, n, γ, δ_1, $\|\bar{c}r^{-\tau_2}\|_{\infty, Q-Q_{\delta_1}}$ and $\|c_2 r^{\beta_2}\|_{\infty, Q-Q_\delta}$. Finally, by Hölder's inequality we have

$$|J_9| \leqq C_{12} \left[\int_{Q_\delta} f^2 r^\theta \Phi^2 \, dx + \int_{Q_\delta} u^2 r^{2-\theta} \Phi^2 \, dx \right]. \tag{3.18}$$

Inserting the estimates (3.10) - (3.18) into (3.9) we infer

(3.19)

$$\begin{aligned}\int_{Q_\delta} |Du|^2 \rho_1 \Phi^2 \, dx \leqq C_{13} \Bigg[&\int_{Q_\delta} u^2 (\tilde{\omega} \circ r)^2 r^{-1} \Phi^2 \, dx \\ &+ \int_{Q_\delta} u^2 r^{-\mu} (\Phi^2 + |D\Phi|^2) \, dx + \sup_{0<s<\delta_0} \int_{\partial Q_s} u^2 \, dS_x + \int_{Q_\delta} f^2 r^\theta \Phi^2 \, dx \Bigg],\end{aligned}$$

where $\mu = \max\{0, 1-2\tau, 2\beta-1, 2\beta_1-1, 1-\tau_1, \tau_1, \beta_1, 1-\tau_2, \beta_2-1, \theta-2\}$. Letting $\delta \to 0$, we conclude $(I) \Rightarrow (II)$ by Lemma 3.2 and Lemma 3.4 and the Lebesgue convergence theorem.

We proceed now to the proof of $(II) \Rightarrow (I)$. We define

$$v(x) = \begin{cases} u(x)(\rho(x) - \delta)\Phi(x)^2 & \text{for} \quad x \in Q_\delta, \\ 0 & \text{elsewhere} . \end{cases}$$

We now observe that we have changed the distance function ρ_1 to $\rho - \delta$. We do this, since the second derivative of ρ_1 behaves like r^{-1} as $\delta \to 0$. On the other hand, the use of $(\rho - \delta)$ in the previous part of the proof may cause some difficulties in estimating J_5'. In fact, $D_j[(\rho - \delta)^{\frac{1}{2}} \rho^\tau]$ does not belong to $L^2(Q_\delta)$. Taking δ_1 as a test function and integrating by parts we obtain

(3.20)

$$\frac{1}{2} \int_{\partial Q_\delta} u^2 A^{ij} D_i\rho D_j\rho \, dS_x = -\frac{1}{2} \int_{\partial Q_\delta} u^2 D_j (A^{ij} D_i \rho \Phi^2) \, dx + \sum_{k=1, k\neq 3}^{9} \widetilde{J}_k,$$

where $\widetilde{J}_k$ is J_k with ρ_1 replaced by $\rho - \delta$. It follows from the Sobolev inequality that

$$|\widetilde{J}_5| \leqq \int_{Q_\delta} |(e_1^i + e_2^i) D_i u u| r \Phi^2 \, dx$$

$$\leqq C_{14} \Big[\int_{Q_\delta} |Du|^2 r \, dx + \int_{Q_\delta} u^2 r^{-\mu} (\Phi^2 + |D\Phi|^2) \, dx \Big],$$

where C_{14} depends on n, δ_1, τ, M_1, γ, $\|e_1^i r^{-\tau}\|_{n, Q - Q_{\delta_1}}$ and $\|e_2^i r^\beta\|_{\infty, Q - Q_{\delta_1}}$ $(i = 1, ..., n)$. Repeating the argument from the previous step, we obtain

$$\int_{\partial Q_\delta} |u|^2 \, dS_x \leqq C_{15} \Big[\int_{Q_\delta - Q_{\delta_1}} u^2 \Big((\tilde{\omega} \circ r)^2 r^{-1} + r^{-\mu} \Big) \, dx + \int_Q |Du|^2 r \, dx + \int_Q f^2 r^\theta \, dx \Big], \tag{3.21}$$

for all $\delta \in (0, \delta_1)$. We now observe that there is some $\delta_2 \in (0, \delta_1)$ such that

$$C_{15} \int_0^{\delta_2} \Big(\tilde{\omega}(t) t^{-1} + t^{-\mu} \Big) \, dt \leqq \frac{1}{2} \tag{3.22.}$$

From (3.21) and (3.22) we infer that

$$\int_{\partial Q_\delta} u^2 \, dS_x \leqq \frac{1}{2} \sup_{\delta < s < \delta_2} \int_{\partial Q_s} u^2 \, dS_x + C_{16}, \tag{3.23}$$

where $\delta \in (0, \delta_2)$ and $C_{16} > 0$ is a constant independent of δ. We conclude from (3.23) that

$$\sup_{\delta_3 < s < \delta_2} \int_{\partial Q_s} u^2 \, dS_x \leqq 2 C_{16}$$

for all $\delta_3 \in (0, \delta_2)$. The result follows by letting $\delta_3 \to 0$.

Let $u \in W^{1,2}_{\text{loc}}(Q)$ be a solution of (1.1) and let us consider the subset $\{u(x_\delta(\cdot));\ 0 < \delta < \delta_\circ\}$ of $L^2(\partial Q)$. By Lemma 1.2 we have

$$\int_{\partial Q} u(x_\delta(x))^2 \, dS_x - \int_{\partial Q_\delta} u(x)^2 dS_\delta = \int_{\partial Q} u(x_\delta(x))^2 \Big[1 - \frac{dS_\delta}{dS_\circ} \Big] \, dS_\circ,$$

with $\frac{dS_\delta}{dS_\circ} \to 1$ uniformly as $\delta \to 0$. Consequently, if (I) holds then $\{u(x_\delta);\ \ 0 < \delta < \delta_\circ\}$ is a bounded subset in $L^2(\partial Q)$. Since bounded sets in $L^2(\partial Q)$ are weakly compact, there exists a sequence $\delta_m \to 0$, as $m \to \infty$, and a function $\zeta \in L^2(\partial Q)$ such that

$$\lim_{\delta_m \to \infty} \int_{\partial Q} u(x_{\delta_m}) g(x) \, dS_x = \int_{\partial Q} \zeta(x) g(x) \, dS_x$$

for each $g \in L^2(\partial Q)$. We now improve this observation by showing that the weak limit of $u \circ x_\delta$ exists.

Theorem 3.2. *If $u \in W^{1,2}_{\rm loc}(Q)$ is a solution of (1.1) such that (I) or (II) holds, then there exists a function $\zeta \in L^2(\partial Q)$ such that $\lim_{\delta\to 0} u \circ x_\delta = \zeta$ weakly in $L^2(\partial Q)$.*

PROOF. Since $A^{ij}D_i\rho D_j\rho$ is uniformly continuous on $\bar{Q}$ with $A^{ij}D_i\rho D_j\rho \geqq \gamma$ on ∂Q (see Lemma 3.4(4)) and the elements of $C^1(\bar{Q})$ restricted to ∂Q are dense in $L^2(\partial Q)$, it suffices to show that

$$\int_{\partial Q_\delta} u(x)g(x)A^{ij}D_i\rho(x)D_j\rho(x)\,dS_x$$

is continuous on $[0,\delta_\circ]$ for each $g \in C^1(\bar{Q})$.
Taking

$$v(x) = \begin{cases} g(x)(\rho(x)-\delta) & \text{for} \quad x \in Q_\delta, \\ 0 & \text{for } x \in Q - Q_\delta, \end{cases}$$

as test function, we obtain

(3.24)

$$\begin{aligned} &\int_{\partial Q_\delta} ugA^{ij}D_i\rho D_j\rho\,dS_x \\ &= \int_{Q_\delta}[a^{ij}-A^{ij}]D_iuD_j\rho g\,dx - \int_{Q_\delta} uD_j\Big(A^{ij}D_i\rho g\Big)\,dx \\ &+ \int_{Q_\delta} a^{ij}D_ju(\rho-\delta)D_ig\,dx + \int_{Q_\delta} b^iuD_i\Big((\rho-\delta)g\Big)\,dx \\ &+ \int_{Q_\delta} d^iD_iu(\rho-\delta)g\,dx + \int_{Q_\delta} cu(\rho-\delta)g\,dx - \int_{Q_\delta} fg(\rho-\delta)\,dx. \end{aligned}$$

It suffices to show that the integrals $I_1, ..., I_7$ on the right-hand side of (3.23) are bounded independently of δ. In what follows, we denote by $C_1, C_2, ...$ positive constants, which are independent of δ and u. We restrict our attention to I_1, I_4 and I_5. By Lemma 3.4(1) we have

$$\begin{aligned} |I_1| &\leqq C_1\int_{Q_\delta}|Du|(\tilde{\omega}\circ r)\,dx \leqq C_1\Big[\int_{Q_\delta - Q_{\delta_\circ}}|Du|(\tilde{\omega}\circ r)\,dx \\ &+ \int_{Q_{\delta_\circ}}|Du|(\tilde{\omega}\circ r)\,dx\Big] \leqq C_2\Big[\int_0^{\delta_\circ}\frac{\tilde{\omega}(t)}{t}\,dt + \int_Q |Du|^2r\,dx\Big]. \end{aligned}$$

It follows from the assumption (A_3) and the Sobolev inequality that

$$\begin{aligned} |I_4| &\leqq C_3\sum_{k=1}^n\int_{Q_\delta}\Big[|b_1^iu| + |b_2^iu|\Big]\,dx \\ &\leqq C_4\sum_{i=1}^n\Big[\|b_1^ir^{-\tau_1}\|_{n,Q}\|r^{\frac{1}{2}(\tau_1-1)}\|_{2,Q}\|D\Big(ur^{\frac{1}{2}(\tau_1+1)}\Big)\|_{2,Q} \\ &+ \|b_2^ir^{\beta_1}\|_{\infty,Q}\|ur^{-\beta_1}\|_{1,Q}\Big] \leqq C_4\Big[\int_Q|Du|^2r\,dx + \int_Q u^2(r^{\tau_1-1}+r^{-\beta_1})\,dx + 1\Big]. \end{aligned}$$

For I_5 we use the decomposition $d^i = e_1^i + e_2^i - b_1^i - b_2^i$ (see (A_3)) and the Hölder inequality to get

$$|I_5| \leqq C_5 \Big[\sum_{i=1}^{n} \|b_1^i \rho^{-\tau_1} + e_1^i \rho^{-\tau}\|_n \|Du(r^{\frac{1}{2}+\frac{\tau_1}{2}} + r^{\frac{1}{2}+\frac{\tau}{2}})\|_2 \|g\|_{2^*}$$

$$+ \|Dur^{\frac{1}{2}}\|_2 \|r^{\frac{1}{2}-\beta_1} + r^{\frac{1}{2}-\beta}\|_2 \Big]$$

and this proves our claim.

3.3. Traces in $L^2(\partial Q)$.

The main objective of this section is to show that a solution u of (1.1) satisfying one of the conditions (I) or (II) has a trace on ∂Q belonging to $L^2(\partial Q)$. To do this we first show that norm of $u(x_\delta)$ converges to norm of ζ. The result then follows from the uniform convexity of the space $L^2(\partial Q)$.

In order to prove the convergence of the norm we use the following function x^δ and technical lemmas.

For $\delta \in (0, \delta_\circ]$ we define the mapping $x^\delta : \bar{Q} \to \bar{Q}_{\frac{\delta}{2}}$ by

$$x^\delta(x) = \begin{cases} x & \text{for} \quad x \in Q_\delta, \\ y_\delta(x) + \frac{1}{2}(x - y_\delta(x)) & \text{for} \quad x \in Q - Q_\delta, \end{cases}$$

where $y_\delta(x)$ is the closest point on ∂Q_δ to x. Thus $x^\delta = x$ for each $x \in Q_\delta$ and $x^\delta = x_{\frac{\delta}{2}}(x)$ for each $x \in \partial Q$. Moreover, $\rho(x^\delta) \geqq \frac{\delta}{2}$ and x^δ is uniformly Lipschitz continuous. Note that if $u \in W^{1,2}_{\text{loc}}(Q)$ then $u(x^\delta) \in W^{1,2}(Q)$.

We begin with the following obvious result (see also Lemma 3.5).

Lemma 3.6. *Let $g \in L^1(Q)$. If $\int_{\partial Q_\delta} |g(x)|\, dS_x$ is bounded on $[0, \delta_\circ]$, then*

$$\frac{1}{\delta} \int_{Q-Q_\delta} |g(x)|\, dS_x \leqq \sup_{0<t<\delta_\circ} \int_{\partial Q_t} |g(x)|\, dS_x.$$

Lemma 3.7. *Let f be a non-negative function in $L^1(Q_{\frac{\delta}{2}} - Q_\delta)$, then*

$$\int_{Q-Q_\delta} f(x^\delta(x))\, dx \leqq \gamma_\circ^4 \int_{Q_{\frac{\delta}{2}}-Q_\delta} f(x)\, dx. \tag{3.25}$$

If $\int_{\partial Q_\delta} f(x)\, dS_x$ is bounded on $[0, \delta_\circ]$ and ω is the Dini function from assumption (A_2), then

$$\int_{Q-Q_\delta} \frac{f(x^\delta(x))\omega(\rho(x))^2}{\rho(x)}\, dx \leqq \gamma_\circ^4 \int_0^\delta \frac{\omega(t)^2}{t}\, dt \sup_{0<t<\delta_\circ} \int_{\partial Q_t} f(x)\, dS_x. \tag{3.26}$$

PROOF. Denoting by J the integral on the left-hand side of (3.26) we obtain

$$J = \int_{x^\delta(Q-Q_\delta)} \frac{f(x)}{2(\rho-\delta)}\omega(2\rho-\delta)^2 J_x^{-1}(x)\,dx \leqq \gamma_\circ^4 \int_{Q_{\frac{\delta}{2}}-Q_\delta} \frac{f(x)}{(2\rho-\delta)}\omega(2\rho-\delta)\,dx$$

$$= \gamma_\circ^4 \int_{\frac{\delta}{2}}^{\delta} \frac{\omega(2t-\delta)^2}{(2t-\delta)} \int_{\partial Q_\delta} f(x)\,dS_x \leqq \gamma_\circ^4 \int_0^\delta \frac{\omega(t)^2}{t}\,dt \sup_{0<t<\delta_\circ} \int_{\partial Q_t} f(x)\,dS_x.$$

Setting $\omega(t) = \sqrt{t}$ we obtain (3.25).

Lemma 3.8. *Let $g \in L^2(Q)$, $\rho^{\frac{1}{2}} f \in L^2(Q)$ and suppose that $\int_{\partial Q_\delta} |g(x)|^2\,dS_x$ is bounded on $(0,\delta_\circ]$, then*

$$\lim_{\delta\to 0} \int_{Q-Q_\delta} f(x^\delta(x))g(x)\,dx = 0.$$

PROOF. Since $\rho(x^\delta) \geqq \frac{\delta}{2}$, we have by Hölder's inequality

$$\left|\int_{Q-Q_\delta} f(x^\delta)g(x)\,dx\right| \leqq \int_{Q-Q_\delta} |f(x^\delta)g(x)|\rho(x^\delta)^{\frac{1}{2}} \frac{\sqrt{2}}{\sqrt{\delta}}\,dx$$

$$\leqq \left[\int_{Q-Q_\delta} f(x^\delta)^2\rho(x^\delta)\,dx\right]^{\frac{1}{2}} \left[\int_{Q-Q_\delta} 2g(x)^2\delta^{-1}\,dx\right]^{\frac{1}{2}}.$$

The result follows from Lemmas 3.6 and 3.7 and the fact that $\lim_{\delta\to 0}|Q-Q_\delta| = 0$.

Lemma 3.9. *If $\rho^{\frac{\alpha}{2}} f \in L^2(Q)$, $0 \leqq \alpha < 1$, $g \in L^2(Q)$ and $\int_{\partial Q_\delta} g(x)^2\,dS_x$ is bounded on $(0,\delta_\circ]$, then*

$$\lim_{\delta\to 0} \int_{Q-Q_\delta} g(x^\delta(x))f(x)\,dx = 0.$$

PROOF. Applying Hölder's inequality and Lemma 3.7 we have

$$\int_{Q-Q_\delta} |g(x^\delta)f(x)|\,dx \leqq \left[\int_{Q-Q_\delta} \frac{g(x^\delta)^2}{\rho(x^\delta)^\alpha}\,dx\right]^{\frac{1}{2}} \left[\int_{Q-Q_\delta} \rho(x)^\alpha f(x)^2\,dx\right]^{\frac{1}{2}}$$

$$\leqq \gamma_\circ^2 \left[\int_{Q_{\frac{\delta}{2}}-Q_\delta} \frac{g(x)^2}{\rho(x)^\alpha}\,dx\right]^{\frac{1}{2}} \left[\int_{Q-Q_\delta} \rho(x)^\alpha f(x)^2\,dx\right]^{\frac{1}{2}}$$

and the result easily follows.

Lemma 3.10. *If $\rho^{\frac{1}{2}} f$ and $\rho^{\frac{1}{2}} g$ belong to $L^2(Q)$, then*

$$\lim_{\delta\to 0} \int_{Q-Q_\delta} f(x^\delta(x))g(x)\rho(x)\,dx = 0.$$

PROOF. Since $\rho(x^\delta(x)) \geqq \rho(x)$, we have

$$\int_{Q-Q_\delta} f(x^\delta)g(x)\rho(x)\,dx \leqq \left[\int_{Q-Q_\delta} f(x^\delta)^2\rho(x^\delta)^2\,dx\right]^{\frac{1}{2}} \left[\int_{Q-Q_\delta} g(x)^2\rho(x)\,dx\right]^{\frac{1}{2}}$$

and the result follows from the previous lemmas.

We are now in a position to prove the main result of this section.

Theorem 3.3. *Let $u \in W^{1,2}_{\mathrm{loc}}(Q)$ be a solution of (1.1) such that one of the conditions (I) or (II) holds. Then there is a function ζ belonging to $L^2(\partial Q)$ such that $u(x^\delta)$ converges to ζ in $L^2(\partial Q)$.*

PROOF. The proof is similar to that of Theorem 2.3 although more care is needed to estimate the integrals involved. First we observe that by Lemma 3.4(4) the norms on $L^2(\partial Q)$ defined by

$$\left(\int_{\partial Q} g(x)^2 A^{ij}(x) D_i\rho(x) D_j\rho(x)\, dS_x\right)^{\frac{1}{2}} \quad \text{and} \quad \left(\int_{\partial Q} g(x)^2\, dS_x\right)^{\frac{1}{2}}$$

are equivalent. We shall prove the convergence of the norm of $u \circ x^\delta$ with respect to the first norm.
Let $v \in W^{1,2}(Q)$. As in Theorem 3.2 we find that

$$\begin{aligned}\int_{\partial Q} \zeta A^{ij} D_i\rho D_j\rho v\, dS_x = \int_Q \Big(&-D_i(A^{ij} v D_j\rho)u + [a^{ij} - A^{ij}] D_i u v D_j\rho \\ &+ a^{ij} D_i u D_j v\rho + (d^i + b^i) D_i u v\rho + b^i u D_i v\rho \\ &+ b^i u v D_i\rho + c u v\rho - f v\rho\Big)\, dx \equiv \int_Q F(v)\, dx.\end{aligned}$$

Since $u(x^\delta) \in W^{1,2}(Q)$, we have

$$\int_{\partial Q} u(x^\delta)\zeta A^{ij} D_i\rho D_j\rho\, dx = \int_{Q-Q_\delta} F(u(x^\delta(x)))\, dx + \int_{Q_\delta} F(u(x))\, dx$$

for $0 < \delta < \delta_0$. We show that

$$\lim_{\delta\to 0} \int_{Q-Q_\delta} F(u(x^\delta(x)))\, dx = 0 \tag{3.27}$$

and that

$$\lim_{\delta\to 0} \int_{Q_\delta} F(u(x))\, dx = \lim_{\delta\to 0} \int_{\partial Q} u(x^\delta)^2 A^{ij}(x) D_i\rho D_j\rho\, dx, \tag{3.28}$$

so that

$$\begin{aligned}\int_{\partial Q} \zeta(x)^2 A^{ij}(x) D_i\rho(x) D_j\rho(x)\, dS_x &= \\ \lim_{\delta\to 0} \int_{\partial Q} \zeta(x) u(x^\delta) A^{ij}(x) D_i\rho(x) D_j\rho(x)\, dS_x& \\ &= \lim_{\delta\to 0} \int_{\partial Q} u(x^\delta)^2 A^{ij}(x) D_i\rho(x) D_j\rho(x)\, dS_x,\end{aligned}$$

since $x^\delta(x) = x_{\frac{\delta}{2}}(x)$ on ∂Q. Consequently the result follows from the uniform convexity of $L^2(\partial Q)$. To prove (3.27) we first observe that

(3.29)

$$
\begin{aligned}
&|F(u(x^\delta))| \leqq C\Big[\tilde{\omega}(r(x))r(x)^{-1}|u(x^\delta)||u(x)| \\
&+ |Du(x^\delta)||u(x)| + |u(x^\delta)||u(x)| + \tilde{\omega}(r(x))|Du(x)||u(x^\delta)| + |Du(x)||Du(x^\delta)|\rho(x) \\
&+ |b+d||Du(x)||u(x^\delta)|\rho(x) + |b||u(x)||Du(x^\delta)|\rho(x) \\
&+ |b||u(x)||u(x^\delta)| + |c||u(x)||u(x^\delta)|\rho(x) + |f||u(x^\delta)|\rho(x)\Big]
\end{aligned}
$$

for $x \in Q - Q_\delta$, where $C > 0$ is a constant independent of δ and u. Now one can show, using Lemmas 3.6 - 3.10 that the integrals of all terms on the right-hand side of the inequality (2.8) over $Q - Q_\delta$ converge to 0 as $\delta \to 0$. We only restrict our attention the integrals involving the first, fourth and sixth term. We denote the corresponding integrals by J_1, J_4 and J_6, respectively.
Applying Young's inequality we obtain

$$
|J_1| \leqq \int_{Q-Q_\delta} \frac{\tilde{\omega}(r)}{r} u(x^\delta)^2\, dx + \int_{Q-Q_\delta} \frac{\tilde{\omega}(r)}{r} u(x)^2\, dx
$$

and by Lemmas 3.4(3), 3.5 and 3.7 $\lim_{\delta\to 0} J_1 = 0$. Similarly

$$
|J_4| \leqq \left[\int_{Q-Q_\delta} \frac{\tilde{\omega}(r)^2}{r} u(x^\delta)^2\, dx\right]^{\frac{1}{2}} \left[\int_{Q-Q_\delta} |Du(x)|^2 r(x)\, dx\right]^{\frac{1}{2}},
$$

and Lemma 3.7 implies that $\lim_{\delta\to 0} J_2 = 0$. Finally, the decomposition for $d^i + b^i = e_1^i + e_2^i$ (see assumption (A_3)) combined with the Sobolev inequality yields

$$
\begin{aligned}
|J_6| &\leqq \sum_{i=1}^n \int_{Q-Q_\delta} |e_1^i||Du(x)||u(x^\delta)|\rho(x)\, dx + \sum_{i=1}^n \int_{Q-Q_\delta} |e_2^i||Du(x)||u(x^\delta)|\rho(x)\, dx \\
&= J_6' + J_6''.
\end{aligned}
$$

Applying the Sobolev inequality we get

$$
\begin{aligned}
|J_6'| &\leqq \sum_{i=1}^n \|e_1^i r^{-\tau}\|_n \left[\int_{Q-Q_\delta} |Du|^2 \rho\, dx\right]^{\frac{1}{2}} \left[\int_{Q-Q_\delta} |u(x^\delta)|^{2^*} \rho(x)^{2^*(\frac{1}{2}+\tau)}\, dx\right]^{\frac{1}{2}} \\
&\leqq C_1 \sum_{i=1}^n \|e_1^i r^{-\tau}\|_n \left(\int_{Q-Q_\delta} |Du|^2 \rho\, dx\right)^{\frac{1}{2}} \left[\left(\int_{Q-Q_\delta} |Du(x^\delta)|^2 \rho\, dx\right)^{\frac{1}{2}} \right. \\
&\left. + \left(\int_{Q-Q_\delta} |u(x^\delta)|^2 \rho^{-1+2\tau}\, dx\right)^{\frac{1}{2}}\right]
\end{aligned}
$$

for some $C_1 > 0$ and consequently, by Lemmas 3.7 and 3.10 $\lim_{\delta\to 0} J_6' = 0$. For J_6'' we have estimate

$$J_6'' \leqq C_2 \int_{Q-Q_\delta} |Du(x)|\rho(x)^{1-\beta}|u(x^\delta)|\, dx$$
$$\leqq C_2\left[\int_{Q-Q_\delta} |Du(x)|^2\rho(x)\, dx + \int_{Q-Q_\delta} |u(x^\delta)|^2\rho(x)^{1-2\beta}\, dx\right]$$

and Lemma 3.7 yields $\lim_{\delta\to 0} J_6'' = 0$.
To show (3.28) we first observe that

$$\int_{Q_\delta}\Big[a^{ij}D_iuD_ju(\rho-\delta) + (a^{ij}-A^{ij})D_iuuD_j\rho$$
$$+ (b^i+d^i)D_iuu(\rho-\delta) + b^iu^2D_i\rho + cu^2(\rho-\delta)$$
$$- fu(\rho-\delta)\Big]\, dx = -\int_{Q_\delta} A^{ij}D_iuuD_j\rho\, dx$$

Combining this identity with the formula for $F(u)$ we find out that

$$\lim_{\delta\to 0}\int_{Q_\delta} F(u)\, dx = \lim_{\delta\to 0}\int_{Q_\delta}\Big[-D_i(A^{ij}D_j\rho u)u - A^{ij}D_iuD_j\rho u\Big]\, dx$$
$$= -\lim_{\delta\to 0}\int_{Q_\delta} D_i(A^{ij}D_j\rho u^2)\, dx = \lim_{\delta\to 0}\int_{\partial Q_\delta} u^2A^{ij}D_i\rho D_j\rho\, dS_x$$
$$= \lim_{\delta\to 0}\int_{\partial Q} u(x_\delta)^2A^{ij}(x)D_i\rho(x)D_j\rho(x)\, dS_x$$
$$+ \lim_{\delta\to 0}\int_{\partial Q} u(x_\delta)^2\Big[A^{ij}(x_\delta)D_i\rho(x_\delta)D_j\rho(x_\delta) - A^{ij}(x)D_i\rho(x)D_j\rho(x)\Big]\, dS_x$$
$$= \lim_{\delta\to 0}\int_{\partial Q} u(x_\delta)^2A^{ij}(x)D_i\rho(x)D_j\rho(x)\, dS_x.$$

3.4. Energy estimate.

The results of Section 3.3 suggest the following approach to the Dirichlet problem with a boundary data in $L^2(\partial Q)$.

Let $\varphi \in L^2(\partial Q)$. A weak solution $u \in W^{1,2}_{\text{loc}}(Q)$ of (1.1) is a solution of the Dirichlet problem with the boundary condition $u(x) = \varphi(x)$ on ∂Q if

$$\lim_{\delta\to 0}\int_{\partial Q}[u(x_\delta(x)) - \varphi(x)]^2 dS_x = 0.$$

It is clear from Theorems 3.1 and 3.3, that if this problem has a solution u then $u \in \widetilde{W}^{1,2}(Q)$. (The definition of this space and its basic properties have been described in Section 1.4.)

In order to prove the existence result we need an energy estimate for the problem depending on a parameter λ

(3.30) $$Lu + \lambda u = f \quad \text{in} \quad Q, \quad u = \varphi \quad \text{on } \partial Q.$$

Theorem 3.4. *There are positive constants $\lambda_\circ$, C and d such that any solution $u \in W^{1,2}_{\mathrm{loc}}(Q)$ of (3.30), with $\lambda > \lambda_\circ$, satisfies*

(3.31)
$$\int_Q |Du(x)|^2 r(x)\,dx + (\lambda - \lambda_\circ)\int_Q u(x)^2 r(x)\,dx$$
$$+ \sup_{0<s<d} \int_{\partial Q_s} u(x)^2 dS_x \leqq C\Big[\int_Q f(x)^2 r(x)^\theta\,dx + \int_{\partial Q} \varphi(x)^2\,dS_x\Big].$$

PROOF. The use of the test function v given by $v(x) = u(x)(\rho(x) - \delta)$ on Q_δ and $v(x) = 0$ on $Q - Q_\delta$ leads to the estimate

(3.32)
$$\gamma\int_\delta |Du(x)|^2(\rho(x) - \delta)\,dx + \lambda\int_{Q_\delta} u(x)^2(\rho(x) - \delta)\,dx$$
$$\leqq \Big[\frac{1}{2}\int_{\partial Q_\delta} A^{ij}u^2 D_i\rho D_j\rho\,dS_x + \int_{Q_\delta} |a^{ij} - A^{ij}||D_i u u D_j\rho|\,dx$$
$$+ \frac{1}{2}\int_{Q_\delta} u^2 D_j(A^{ij}D_i\rho)\,dx + \int_{Q_\delta} |b^i u||D_i\Big((\rho - \delta)u\Big)|\,dx$$
$$+ \int_{Q_\delta} |d^i D_i u u|(\rho - \delta)\,dx + \int_{Q_\delta} |c|u^2(\rho - \delta)\,dx + \int_{Q_\delta} |fu|(\rho - \delta)\,dx\Big].$$

Repeating the estimates from the proofs of Theorems 3.1 and 3.3 we get, letting $\delta \to 0$,

(3.33)
$$\int_Q |Du|^2(\rho - \delta)\,dx + \lambda\int_Q u^2(\rho - \delta)\,dx$$
$$\leqq C_1\Big[\int_{\partial Q} \varphi^2\,dS_x + \int_Q f^2\rho^\theta\,dx$$
$$+ \int_Q u^2(\rho^{-\mu} + \frac{\tilde{\omega}\circ\rho}{\rho})\,dx + \int_Q u^2\,dx\Big],$$

where $C_1 > 0$ is a constant of the same nature as constants C_i from the proofs of Theorems 3.1 and 3.3 and μ is a constant from the proof of Theorem 3.1. To proceed further let us denote the sum of the last three integrals on the right-hand side of (3.33) by A. Similarly, for $\delta \in (0, \delta_\circ]$ we get

$$\int_{\partial Q_\delta} u^2\,dS_x \leq C_2\Big[\int_Q |Du|^2\rho\,dx + \lambda\int_Q u^2\rho\,dx + A\Big]$$

for some $C_2 > 0$ independent of δ and u. Thus (3.32) and (3.33) yield

$$\int_Q |Du|^2 \rho\, dx + \sup_{0<s<d} \int_{\partial Q_s} u^2\, dS_x + \lambda \int_Q u^2 \rho\, dx$$
$$\leqq C_3 \Big[\int_{\partial Q} \varphi^2\, dS_x + A \Big].$$

The proof can now be completed by estimations based on Lemma 3.5. By virtue of (3.6) and (3.7) we obtain

$$\int_Q u(x)^2 r(x)^{-\mu}\, dS_x \leqq \frac{d^{1-\mu}}{1-\mu} \sup_{0<t<d} \int_{\partial Q_t} u(x)^2\, dS_x + \frac{1}{d_1} \int_Q u(x)^2 r(x)\, dx,$$

and

$$\int_Q u(x)^2 \tilde{\omega}(r)^2 r^{-1}\, dx \leqq \tilde{\omega}(d) \int_0^d \frac{\tilde{\omega}(t)}{t}\, dt \sup_{0<t<d} \int_{\partial Q_t} u(x)^2\, dS_x$$
$$+ \frac{\sup \tilde{\omega}^2}{d_1^2} \int_Q u(x)^2 r(x)\, dx,$$

where $d_1 = \inf_{Q_d} r(x)$. The conclusion now follows by choosing λ sufficiently large and d sufficiently small.

Remark. It is clear that Theorems 3.1 - 3.4 remain true with f replaced by $f - D_i g^i$, where $g^i \in L^2(Q)$. Obviously the right-hand side of the energy estimate is modified by adding the term $\int_Q |g|^2\, dx$.

3.5. Solvability of the Dirichlet problem.

The proof of the existence result for a bounded domain is patterned after the proof of Theorem 2.5.

Theorem 3.5. *Let $\varphi \in L^2(\partial Q)$ and let C, $\lambda_\circ$ and d be the constants from Theorem 3.4. If $\lambda > \lambda_\circ$, then the Dirichlet problem (3.30) (with f replaced by $f - D_i g^i$, where $g^i \in L^2(Q)$) has a unique solution $u \in \widetilde{W}^{1,2}(Q)$. Moreover, this solution obeys the estimate (3.31).*

PROOF. We first define the truncation operator

$$T_m(w) = \begin{cases} w(x) & \text{for} \quad x \in Q_{\frac{1}{m}}, \\ 0 & \text{for} \quad x \in Q - Q_{\frac{1}{m}}. \end{cases}$$

By L_m we denote the elliptic operator obtained from L by replacing b^i, d^i and c by $T_m b^i$, $T_m d^i$ and $T_m c$, respectively. If φ is a trace of a function belonging to $W^{1,2}(Q)$, then the problem (3.29) is coercive for λ sufficiently large, say $\lambda \geqq \lambda_1$. Here λ_1 can be chosen independently of m. Consequently the Dirichlet problem is solvable (see [**ST1**] or [**ST2**]). To complete the proof we employ the approximation argument used in the proof of Theorem 2.5.

As in Section 2.4 we observe that if for some constant $c_\circ > 0$

$$-D_i b^i + c \geqq c_\circ \quad \text{on} \quad Q, \tag{3.34}$$

then $\lambda = 0$.

Theorem 3.6. *Let $\varphi \in L^2(\partial Q)$ and and suppose that (3.34) holds. If $e_2^i \equiv 0$, $b_2^i \equiv 0$ $(i = 1, ..., n)$ and $c_2 \equiv 0$ on Q, then the Dirichlet problem*

$$(3.35) \qquad Lu = f - D_i g^i \quad \text{in} \quad Q, \quad u = \varphi \quad \text{on} \quad \partial Q$$

has a unique solution in $\widetilde{W}^{1,2}(Q)$. (Here we assume that g^i are in $L^2(Q)$.

The proof is similar to that of Theorem 2.6 and therefore is omitted.

Finally, we discuss the relation between $\widetilde{W}^{1,2}$–solutions and $W^{1,2}$–solutions of the Dirichlet problem.

Theorem 3.7. *Suppose that $f \in L^2(Q)$ and let $\varphi \in L^2(\partial Q)$. Moreover we assume that $e_2^i \equiv 0$, $b_2^i \equiv 0\,(i = 1, .., n)$ and $c_2 \equiv 0$. If there is a function $\varphi_1 \in W^{1,2}(Q)$ such that $\varphi_1 = \varphi$ on ∂Q in the sense of trace, then a solution $u \in \widetilde{W}^{1,2}(Q)$ of the Dirichlet problem (3.35) is a solution in $W^{1,2}(Q)$ of the same problem.*

PROOF. We first observe that $b_i \in L^n(Q)$ and $c \in L^{\frac{n}{2}}(Q)$. For λ sufficiently large, say $\lambda \geqq \lambda_o$, the Dirichlet problem (3.30) has a unique solution in $W^{1,2}(Q)$ and in $\widetilde{W}^{1,2}(Q)$. Let us denote these solutions by w and u, respectively. Since a solution in $W^{1,2}(Q)$ of the problem (3.30) (with f replaced by $f - D_i g^i$) is also a solution in $\widetilde{W}^{1,2}(Q)$ of the same problem, we see that $w \equiv u$ on Q. In the case of the problem (3.35) we consider the problem in $\widetilde{W}^{1,2}(Q)$

$$Lv + \lambda v = f - D_i g^i + \lambda u \text{ in } Q,\ u = \varphi \text{ on } \partial Q,$$

where $\lambda \geqq \lambda_o$ and u is a solution of (3.30). By the previous part of the proof v is also a solution in $W^{1,2}(Q)$. By the uniqueness we must have $v \equiv u$.

The existence results with a boundary data in $L^p(Q)$i, with $p \geqq 1$ can be found in [CH6], [GM], [HW1], [HW2], [LI2] and [PE1].

CHAPTER 4

ESTIMATES OF DERIVATIVES

In this chapter we derive estimates of the derivatives of the second order. We also establish a higher order integrability property of the derivatives of the first order. The main tool for our treatment of the higher integrability property of Du is a Gehring's inverse maximal function inequality. Finally, we briefly discuss an analogue of Theorem 3.1 for the second order derivatives.

4.1. Estimates of the second order derivatives.

Throughout this section we assume that (A_1) holds, $D_k a^{ij}$ and $D_k b^i$ $(k,i,j = 1,...,n)$ belong to $L^\infty(Q)$. Moreover d^i $(i = 1,...,n)$ and c belong to $L^q(Q)$ for some $q > n$.

It is known that under these assumptions any solution of (1.1) in $W^{1,2}_{\text{loc}}(Q)$ belongs to $W^{2,2}_{\text{loc}}(Q)$ (see **[LU]**).

Theorem 4.1. *If $u \in \widetilde{W}^{1,2}(Q)$ is a solution of (1.1) then*

$$\int_Q |D^2u(x)|^2 r(x)^3\,dx \leqq C\left[\int_Q |Du(x)|^2 r(x)\,dx + \int_Q u(x)^2 r(x)\,dx + \int_Q f(x)^2 r(x)^\theta\,dx\right].$$

PROOF. Let w be a function in $W^{2,2}(Q)$ with compact support in Q. Using $D_k w$ as a test function and integrating by parts we obtain

$$\begin{aligned}(4.1)\qquad &\int_Q D_k a^{ij} D_i u D_j w\,dx + \int_Q a^{ij} D_{ik} u D_j w\,dx \\ &+ \int_Q D_k b^i u D_i w\,dx + \int_Q b^i D_k u D_i w\,dx - \int_Q d^i D_i u D_k w\,dx \\ &- \int_Q cu D_k w\,dx = -\int_Q f D_k w\,dx.\end{aligned}$$

To proceed further we set

$$w(x) = \begin{cases} D_k u(x)(\rho(x) - \delta)^3 & \text{for} \quad x \in Q_\delta, \\ 0 & \text{for} \quad x \in Q - Q_\delta. \end{cases}$$

We then obtain

$$
\begin{aligned}
&\int_{Q_\delta} a^{ij} D_{ik}u D_{jk}u(\rho-\delta)^3\,dx + 3\int_{Q_\delta} a^{ij} D_{ik}u D_k u D_j\rho(\rho-\delta)^2\,dx \\
&+\int_{Q_\delta} D_k a^{ij} D_i u D_{jk}u(\rho-\delta)^3\,dx + 3\int_{Q_\delta} D_k a^{ij} D_i u D_k u D_j\rho(\rho-\delta)^2\,dx \\
&+\int_{Q_\delta} D_k b^i u D_{ik}u(\rho-\delta)^3\,dx + 3\int_{Q_\delta} D_k b^i u D_k u D_i\rho(\rho-\delta)^2\,dx \\
&+\int_{Q_\delta} b^i D_k u D_{ik}u(\rho-\delta)^3\,dx + 3\int_{Q_\delta} b^i D_k u D_k u D_i\rho(\rho-\delta)^2\,dx \\
&-\int_{Q_\delta} d^i D_i u D_{kk}u(\rho-\delta)^3\,dx - 3\int_{Q_\delta} d^i D_i u D_k u D_k\rho(\rho-\delta)^2\,dx \\
&-\int_{Q_\delta} cu D_{kk}u(\rho-\delta)^3\,dx - 3\int_{Q_\delta} cu D_k u D_k\rho(\rho-\delta)^2\,dx \\
&= -\int_{Q_\delta} f D_{kk}u(\rho-\delta)^3\,dx - 3\int_{Q_\delta} f D_k u D_k\rho(\rho-\delta)^2\,dx.
\end{aligned}
$$

It follows from (A_1) and the boundedness of b^i, $D_k b^i$, a^{ij} and $D_k a^{ij}$ and the Young inequality that

$$
\begin{aligned}
&\int_{Q_\delta} |Du|^2(\rho-\delta)^3\,dx \leqq C_1\Big[\int_{Q_\delta} |Du|^2(\rho-\delta)\,dx \\
&+\int_{Q_\delta} u^2(\rho-\delta)\,dx + \int_{Q_\delta} |d^i||D_i u||D_{kk}u|(\rho-\delta)^3\,dx \\
&+3\int_{Q_\delta} |d^i||D_i u||D_k u|(\rho-\delta)^2\,dx + \int_{Q_\delta} |c||u||D_{ik}u|(\rho-\delta)^3\,dx \\
&+3\int_{Q_\delta} |c||D_k u||u|(\rho-\delta)^2\,dx + \int_{Q_\delta} f^2(\rho-\delta)^\theta\,dx\Big].
\end{aligned}
$$

Now a routine application of the Sobolev inequality to the integrals containing c and d^i yields the desired estimate.

The following improvement of the energy estimate for the problem (3.30) follows immediately from Theorems 3.4 and 4.1.

Theorem 4.2. *There are positive constants C, $\lambda_\circ$ and d, such that any solution $u \in W^{2,2}_{\mathrm{loc}}(Q)$ of the problem (3.30), with $\lambda > \lambda_\circ$, satisfies*

$$
\begin{aligned}
&\int_Q |D^2u(x)|^2 r(x)^3\,dx + \int_Q |Du(x)|^2 r(x)\,dx \\
&+(\lambda-\lambda_\circ)\int_Q u(x)^2 r(x)\,dx + \sup_{0<s<d}\int_{\partial Q_s} u(x)^2\,dS_x \\
&\leqq C\Big[\int_Q f(x)^2 r(x)^\theta\,dx + \int_{\partial Q} \varphi(x)^2\,dS_x\Big].
\end{aligned}
$$

4.2. Analogue of Theorem 3.1 for Du.

In this section we briefly discuss the behaviour of surface integrals of $|Du|^2$. We prove the analogue of Theorem 3.1. As details of the proof are similar to those of Theorem 3.1 we only sketch the proof.

Theorem 4.3. *Suppose that* $\int_Q f(x)^2 r(x)^\mu\, dx < \infty$, $0 \leqq \mu < 1$. *Let* $u \in W^{1,2}_{\text{loc}}(Q)$ *be a solution of (3.1). Then the following conditions*

$$\text{(I)} \qquad \sup_{0<\delta<\delta_0} \int_{\partial Q_\delta} |Du(x)|^2\, dS_x < \infty$$

and

$$\text{(II)} \qquad \int_Q \sum_{i,j=1}^n |D_{ij}u(x)|^2 r(x)\, dx < \infty$$

are equivalent.

PROOF. Since our hypotheses on coefficients are more restrictive than in Theorem 3.1, we may use the same test function

$$w(x) = \begin{cases} D_k u(x)(\rho(x) - \delta) & \text{for} \quad x \in Q_\delta, \\ 0 & \text{for} \quad x \in Q - Q_\delta \end{cases}$$

to prove both implications $(I) \Rightarrow (II)$ and $(I) \Rightarrow (II)$. We derive on substitution from (4.1)

$$\begin{aligned} &\int_{Q_\delta} \Big[a^{ij} D_{ik}u D_{jk}u(\rho - \delta) + a^{ij} D_{ik}u D_k u D_j\rho + D_k a^{ij} D_i u D_{jk}u(\rho - \delta) \\ &+ D_k a^{ij} D_i u D_k u D_j \rho + D_k b^i u D_{ik}u(\rho - \delta) + D_k b^i u D_k u D_i\rho + b^i D_k u D_i u(\rho - \delta) \\ &+ b^i D_k u D_k u D_i \rho - d^i D_i u D_{kk}u(\rho - \delta) - d^i D_i u D_k u D_k\rho - cu D_{kk}u(\rho - \delta) \\ &- cu D_k u D_k \rho \Big]\, dx = - \int_{Q_\delta} \Big[f D_{kk}u(\rho - \delta) - f D_k u D_k \rho \Big]\, dx. \end{aligned}$$

To prove $(I) \Rightarrow (II)$ we integrate the integral of the second term by parts to get

$$\begin{aligned} &\int_{Q_\delta} |D(D_k u)|^2 (\rho - \delta)\, dx \leqq C \Big[\int_{\partial Q_\delta} |D_k u|^2\, dS_x \\ &+ \int_{Q_\delta} \Big(|Du|^2 + |Du|^2(\rho - \delta) + u^2(\rho - \delta) + u^2 \\ &+ |d^i||D_i u||D_k u| + |d^i||D_i u||D_{kk}u|(\rho - \delta) + |c||u||D_{kk}u|(\rho - \delta) \\ &+ |c||u||D_k u| + f^2(\rho - \delta)^\mu + |D_k u|(\rho - \delta)^{-\mu} \Big)\, dx \Big] \end{aligned}$$

for a suitable positive constant C independent of δ and u. Since $D_k u \in W^{1,2}_{\mathrm{loc}}(Q)$ and (I) holds, it follows from Lemma 1.4 that

$$\int_{Q_\delta} |Du(x)|^2 (\rho(x) - \delta)^{-\beta}\, dx \tag{4.2}$$

is bounded as a function of $\delta \in (0, \delta_\circ]$ for every $0 \leqq \beta < 1$. On the other hand Theorem 3.1 implies that $u \in L^2(Q)$. The result now follows repeating the argument of the corresponding part of Theorem 3.1.
In order to prove $(II) \Rightarrow (I)$ we note that, by Lemma 1.5, the relation (4.2) remains true and the proof is similar to the part $(II) \Rightarrow (I)$ of Theorem 3.1.

It is now clear that Theorem 4.3 can be used to show the existence of the limits in $L^2(\partial Q)$ of $D_k u(x_\delta)$ (see Theorems 3.2 and 3.3) but we are not going to develop these ideas here.

4.3. Reverse Hölder inequality.

In this section we prove the reverse Hölder inequality. This result is stated in terms of the Hardy - Littlewood maximal functions.

Given a measurable and bounded set $A \subset R_n$ with $|A| > 0$ we define for $f \in L^1(R_n)$

$$⨍_A f(x)\, dx = \frac{1}{|A|} \int_A f(x)\, dx.$$

A maximal function Mf of $f \in L^1_{\mathrm{loc}}(R_n)$ is defined by

$$Mf(x) = \sup_{r>0} ⨍_{Q(x,r)} |f(y)|\, dy,$$

where $Q(x,r)$ denotes a cube of the diameter $2r$ centered at x and whose edges are parallel to the coordinate axes. It is well known that if $f \in L^p(R_n)$, $1 \leqq p < \infty$, then $Mf(x) < \infty$ a.e. on R_n and $Mf \in L^p(R_n)$ if $1 < p < \infty$ (see [ST]).

Let $F(t)$ be a non-negative, convex and increasing function on $[0, \infty)$ such that $F(0) = 0$ and $F(2t) \leq KF(t)$ for all $t \in [0, \infty)$, where K is a positive constant. We denote by $L_F(A)$ the set of all measurable functions on A such that

$$\|f\|_F = F^{-1}\left(⨍_A F(|f(x)|)\, dx\right) < \infty.$$

We associate with F and $f \in L_{F,\mathrm{loc}}(R_n)$ the following variant of the maximal function

$$M_F(f)(x) = F^{-1}\left(\sup_{r>0} ⨍_{Q(x,r)} F(|f(y)|) dy\right).$$

It is clear that

$$M_F(f) = F^{-1}\Big(M(F(|f|))\Big).$$

It follows from the Jensen inequality that

$$M(f)(x) \leqq M_F(f)(x) \quad \text{for } x \in R_n.$$

Our first objective is to show that the reverse inequality for maximal functions improves the integrability of f.

We commence with the following generalization of Gehring's inequality for Stieltjes integrals due to Iwaniec [IW] .

Lemma 4.1. *Suppose that β is a positive and differentiable function on $[a,\infty)$ with $\beta'(t) > 0$ on $[0,\infty)$ and h is a non-negative and non-increasing function on $[a,\infty)$ with $\lim_{t\to\infty} h(t) = 0$. Further we assume that*

$$-\int_t^\infty \beta(s)\,dh(s) \leqq C\beta(t)h(t) \tag{4.3}$$

for all $t \in [a,\infty)$, where C is a positive constant. Then

$$-\int_a^\infty \beta(t)^p\,dh(t) \leqq \frac{\beta(a)^{p-1}}{C - p(C-1)}\left(-\int_a^\infty \beta(t)\,dh(t)\right) \tag{4.4}$$

for all $p \in [1, \frac{C}{C-1})$.

PROOF. We may assume without loss of generality that $h(t) = 0$ for sufficiently large t. Integrating by parts we obtain

$$\begin{aligned}
&-\int_a^\infty \beta(t)^p\,dh(t) = \int_a^\infty \beta(t)^{p-1}\left(\int_t^\infty \beta(s)\,dh(s)\right)' dt \\
&= -\beta(a)^{p-1}\int_a^\infty \beta(s)\,dh(s) + (p-1)\int_a^\infty \beta(t)^{p-2}\beta'(t)\left(-\int_t^\infty \beta(s)\,dh(s)\right) dt \\
&\leqq -\beta(a)^{p-1}\int_a^\infty \beta(s)\,dh(s) + C(p-1)\int_a^\infty \beta'(t)\beta(t)^{p-1}h(t)\,dt \\
&= -\beta(a)^{p-1}\int_a^\infty \beta(s)\,dh(s) + \frac{C(p-1)}{p}\left(-\int_a^\infty \beta(t)^p\,dh(t)\right) \\
&- \frac{C(p-1)}{p}\beta(a)^p h(a).
\end{aligned}$$

Using (4.3), we then have

$$-\int_a^\infty \beta(t)^p\,dh(t) \leqq \frac{C(p-1)}{p}\left(-\int_a^\infty \beta(t)^p\,dh(t)\right) - \frac{\beta(a)^{p-1}}{p}\int_a^\infty \beta(s)\,dh(s).$$

The last inequality is equivalent to (4.4) and this completes the proof.

The proof of the reverse Hölder inequality theorem is based on the theorem of Zygmund and Calderon.

Theorem 4.4. *Let f be a non-negative function in $L^1(R_n)$ and let α be a positive constant. Then there exists a decomposition of R_n such that*

(1) $R_n = F \cup \Omega$, $F \cap \Omega = \varnothing$,

(2) $f(x) \leqq \alpha$ a.e on F,

(3) *Ω is the union of cubes, $\Omega = \bigcup D_k$, whose interiors are disjoint, and so that for each D_k*

$$\alpha < \frac{1}{|D_k|} \int_{D_k} f(x)\, dx \leqq 2^n \alpha.$$

For the proof we refer to **[ST]** (p.17).

Theorem 4.5. *Suppose that D is a cube in R_n and that $f \in L_F(D)$ with $f(x) \geqq 0$ on D and $\|f\|_F = F^{-1}(⨍_D F(f)\, dx) < \infty$. We extend f to be 0 outside D. Further, we assume that*

$$M_F(f)(x) \leqq \beta M(f)(x) + T \quad \text{a.e. on} \quad D, \tag{4.5}$$

where β and T are constants such that $0 \leqq T$ and $1 \leqq \beta$. Then there exist positive constants $p_\circ = p_\circ(\beta, n, F) > 1$ and $A_p = A_p(\beta, n, F, T, \|f\|_F)$ such that

$$\|f\|_G \leqq A_p \tag{4.6}$$

for all $1 \leqq p < p_\circ$, where $G(t) = \frac{F(t)^p}{t^{p-1}}$.

PROOF. Let $a = \max\{\frac{T}{\beta}, \|f\|_F\}$ and set $s = F^{-1}(2^n F(3t\beta))$ for fixed $t \geqq a$. Thus

$$F(s) \geqq F(a) \geqq ⨍_D F(f)\, dx.$$

It follows from Theorem 4.4 that there exist cubes D_j with disjoint interiors such that

$$F(s) \leqq ⨍_{D_j} F(f)\, dx \leqq 2^n F(s) \tag{4.7}$$

and

$$F(f(x)) \leqq F(s) \quad \text{a.e. on} \quad D - \bigcup_j D_j. \tag{4.8}$$

Obviously (4.8) implies that $f(x) \leqq s$ a.e. on $D - \bigcup_j D_j$. Let us now define

$$E_t = \{x \in D;\quad f(x) > t\}.$$

It then follows from (4.8) that $|E_s - \bigcup D_j| = 0$ and consequently, by (4.7), we obtain

$$\int_{E_s} F(f)\, dx \leqq \sum_j \int_{D_j} F(f)\, dx \leqq 2^n F(s) |\bigcup_j D_j|. \tag{4.9}$$

Our next step is to estimate the measure of $\bigcup D_j$. Observe that if $x \in D_j$ then $D_j \subset Q(x,r)$, where $r = \text{diam } D_j$ and consequently

$$F(M_F(f)(x) = \sup_{r>0} \fint_{Q(x,r)} F(f)\,dx \geqq \frac{1}{2^n} \fint_{D_j} F(f)\,dx > \frac{F(s)}{2^n} = F(3t\beta).$$

Using this, (4.5) and the inequality $t \geqq a \geqq \frac{T}{\beta}$ we obtain

$$M(f)(x) \geqq \beta^{-1} M_F(f)(x) - \beta^{-1}T > 3t - \beta^{-1}T \geqq 2t$$

for a.e. $x \in \bigcup D_j$. According to the definition of Mf, for a.e. $x \in \bigcup D_j$ there is a cube D_x such that

$$\fint_{D_x} f\,dx > 2t.$$

Using the Vitali covering lemma, from this family of cubes $\{D_x\}$ we can extract a disjoint sequence of cubes, which we denote by $\{D_{x_j}\}$ such that

$$|\bigcup_j D_j| \leqq 3^n \sum_i |D_{x_i}| \tag{4.10}$$

and

$$2t|D_{x_j}| \leqq \int_{D_{x_j}} f\,dx \leqq \int_{E_t \cap D_{x_j}} f\,dx + t|D_{x_j}|.$$

Consequently $t\sum_i |D_{x_i}| \leqq \int_{E_t} f\,dx$. This combined with (4.10) leads to the estimate

$$t|\bigcup D_j| \leqq 3^n \int_{E_t} f\,dx.$$

Inserting the last estimate into (4.9) we obtain

$$\int_{E_s} F(f)\,dx \leqq \frac{6^n F(s)}{t} \int_{E_t} f\,dx.$$

From this and the obvious inequality

$$\int_{E_t - E_s} F(f)\,dx = \int_{E_t - E_s} f \frac{F(f)}{f}\,dx \leqq \frac{F(s)}{t} \int_{E_t} f\,dx$$

we deduce that

$$\int_{E_t} F(f)\,dx \leqq \frac{(6^n+1)F(s)}{t} \int_{E_t} f\,dx \leqq C(n,\beta,F)\frac{F(t)}{t} \int_{E_t} f\,dx \tag{4.11}$$

for $t \geqq a$. Let us define the function $h(t) = \int_{E_t} f\,dx$. It is clear that

$$\int_{E_t} F(f)\,dx = -\int_t^\infty \frac{F(s)}{s}\,dh(s) \tag{4.12}$$

and consequently the inequality (4.11) takes the form

$$-\int_t^{\infty} \frac{F(s)}{s}\, dh(s) \leqq C(n,\beta,F)\frac{F(t)}{t} h(t)$$

for $t \geqq a$. On the other hand it follows from Lemma 4.1 that

$$-\int_a^{\infty} \frac{F(s)^p}{s^p}\, dh(s) \leqq \frac{F(a)^{p-1}a^{1-p}}{C-p(C-1)}\left(-\int_a^{\infty} \frac{F(s)}{s}\, dh(s)\right)$$

for $p \in [1, C(C-1)^{-1})$. Again, using (4.12) the last inequality can be written in the form

$$\int_{E_a} \frac{F(f)^p}{f^{p-1}}\, dx \leqq \frac{F(a)^{p-1}a^{1-p}}{C-p(C-1)} \int_{E_a} F(f)\, dx.$$

Using this and obvious estimate

$$\int_{D-E_a} \frac{F(f)^p}{f^{p-1}}\, dx \leqq \frac{F(a)^{p-1}}{a^{p-1}} \int_{D-E_a} F(f)\, dx$$

we obtain the estimate

$$\int_D G(f)\, dx \leqq \frac{G(a)}{F(a)(C-p(C-1))} \int_D F(f)\, dx \leqq \frac{G(a)|D|}{C-p(C-1)}.$$

Finally, using the convexity of F and the inequality $0 < C-p(C-1) \leqq 1$, we arrive at the estimate

$$\|f\|_G \leqq G^{-1}\left(\frac{G(a)}{C-p(C-1)}\right) \leqq \frac{a}{C-p(C-1)} = \frac{\max\{T\beta^{-1}, \|f\|_F\}}{C-p(C-1)}$$

and the inequality (4.5) follows.

4.4. Higher integrability property of Du.

To show an application of Theorem 4.5 we need the following form of the Sobolev inequality

$$\left(\int_Q |u-a|^{n^*}\, dx\right)^{\frac{1}{n^*}} \leqq S\|Du\|_p \tag{4.13}$$

for all $u \in W^{1,p}(Q)$ with $1 \leqq p < n$, where $a = ⨍_Q u\, dx$ and S is a positive constant depending on p and n and $\frac{1}{n^*} = \frac{1}{p} - \frac{1}{n}$. We restrict our attention to the problem

$$D_i\left(a^{ij}(x)D_j u\right) = 0 \quad \text{in} \quad Q, \quad u(x) = \varphi(x) \quad \text{on} \quad \partial Q, \tag{4.14}$$

where the coefficients satisfy (A_1) and (A_2).

Theorem 4.6. *Let $\varphi \in L^2(\partial Q)$ and let $u \in \widetilde{W}^{1,2}(Q)$ be a solution to the problem (4.14). Then there exists a positive constant ϵ such that*

$$\int_Q \left(|Du(x)|^2 r(x)^{n+1}\right)^{1+\epsilon} dx < \infty.$$

PROOF. For a fixed $y \in Q$ we denote by $Q(y,r)$ and $Q(y,2r)$ cubes such that $Q(y,r) \subset Q(y,2r) \subset Q$. Let β be a non-negative function in $C_\circ^1(Q(y,2r))$ such that $\beta(x) = 1$ on $Q(y,r)$, $0 \leqq \beta(x) \leqq 1$ on $Q(y,2r)$ and $|D\beta(x)| \leqq Cr^{-1}$ on $Q(y,2r)$ for some constant $C > 0$. Let us now set

$$v(x) = (u(x) - a)\beta(x)^2,$$

where $a = \fint_{Q(y,2r)} u(x)\, dx$. Taking v as a test function, we obtain

$$\int_Q a^{ij} D_i u D_j u \beta^2\, dx + 2\int_Q a^{ij} D_i u(u-a)\beta D_j \beta\, dx = 0$$

and consequently

$$\int_{Q(y,r)} |Du|^2\, dx \leqq \frac{C_1}{r}\int_{Q(y,2r)} |Du||u-a|\, dx. \tag{4.15}$$

By virtue of (4.13) with $p = \frac{2n}{n+1}$ and the Hölder inequality we get

$$\begin{aligned}\int_{Q(y,2r)} |Du||u-a|\, dx &\leqq \left(\int_{Q(y,2r)} |u-a|^{\frac{2n}{n-1}}\, dx\right)^{\frac{n-1}{2n}} \left(\int_{Q(y,2r)} |Du|^{\frac{2n}{n+1}}\, dx\right)^{\frac{n+1}{2n}} \\ &\leqq S\left(\int_{Q(y,2r)} |Du|^{\frac{2n}{n+1}}\, dx\right)^{\frac{n+1}{n}}.\end{aligned}$$

Inserting this estimate into (4.15) we obtain

$$\fint_{Q(y,r)} |Du|^2\, dx \leqq C_2\left(\fint_{Q(y,2r)} |Du|^{\frac{2n}{n+1}}\, dx\right)^{\frac{n+1}{n}}. \tag{4.16}$$

To proceed further we set

$$f(x) = \left(r(x)^{n+1}|Du(x)|^2\right)^{\frac{n}{n+1}},$$

where $r(x) = 0$ for $x \notin Q$. It is obvious that for any cube $Q(y,r)$ we have

$$\begin{aligned}\fint_{Q(y,r)} f^{\frac{n+1}{n}}\, dx &= \fint_{Q(y,r)} r(x)^{n+1}|Du(x)|^2\, dx \\ &\leqq \sup_{x\in Q(y,r)} r(x)^n \fint_{Q\cap Q(y,r)} r(x)|Du(x)|^2\, dx.\end{aligned}$$

We now distinguish two cases:

$$(i)\quad r(y) \leqq 5r \qquad \text{and} \qquad (ii)\quad r(y) > 5r.$$

In the first case we have $\sup_{x\in Q(y,r)} r(x)^n \leqq (6r)^n$ and consequently

$$\fint_{Q(y,r)} f^{\frac{n+1}{n}}\,dx \leqq \frac{(6r)^n}{|Q(y,r)|}\int_Q r(x)|Du(x)|^2\,dx \leqq C_3(n)\int_Q r(x)|Du(x)|^2\,dx.$$

In the second case we first observe that

$$\sup_{x\in Q(y,r)} r(x) \leqq C_4(n) \inf_{x\in Q(y,2r)} r(x)$$

and therefore by (4.16) we get

$$\begin{aligned}
\fint_{Q(y,r)} f^{\frac{n+1}{n}}\,dx &= \fint_{Q(y,r)} r(x)^{n+1}|Du(x)|^2\,dx \\
&\leqq \sup_{Q(y,r)} r(x)^{n+1} \fint_{Q(y,r)} |Du(x)|^2\,dx \\
&\leqq C_2 \sup_{Q(y,r)} r(x)^{n+1}\Big(\fint_{Q(y,2r)} |Du(x)|^{\frac{2n}{n+1}}\,dx\Big)^{\frac{n+1}{n}} \\
&\leqq C_4(n)C_2\Big[\fint_{Q(y,2r)}\Big(r(x)^{n+1}|Du(x)|^2\Big)^{\frac{n}{n+1}}\,dx\Big]^{\frac{n+1}{n}} \\
&= C_4(n)C_2\Big(\fint_{Q(y,2r)} f\,dx\Big)^{\frac{n+1}{n}}.
\end{aligned}$$

Combining both cases together we get

$$\fint_{Q(x,r)} f^{\frac{n+1}{n}} \leqq C_5\Big(\fint_{Q(y,2r)} f\,dx\Big)^{\frac{n+1}{n}} + T,$$

where $C_2 = C_4(n)C_2$ and

$$T = C_3(n)\int_Q r(x)|Du(x)|^2\,dx.$$

Applying Theorem 4.5 with $F(t) = t^{\frac{n+1}{n}}$ the result follows.

In the next chapter we shall need the following form of a higher integrability property.

Theorem 4.7. *Let $u \in W^{1,2}(Q)$ be a solution of*

$$-D_i(a^{ij}(x)D_j u) = f(x) - D_i g^i(x) \quad \text{in} \quad Q$$

with $f \in L^\sigma(Q)$ and $g^i \in L^r(Q)$, $i = 1, ..., n$, for some $r > 2$ and $\sigma > 2$. Then there exists $p > 2$ such that $u \in W^{1,p}_{loc}(Q)$. Moreover, for $B(y, \frac{R}{2}) \subset B(y, R) \subset Q$ we have

$$\left(\fint_{B(y,\frac{R}{2})} (u^2 + |Du|^2)^{\frac{p}{2}}\, dx\right)^{\frac{1}{p}} \leqq C\left(\fint_{B(y,R)} (u^2 + |Du|^2)\, dx\right)^{\frac{1}{2}}$$

$$+ \left(\fint_{B(y,R)} (f^2 + |g|^2)^{\frac{p}{2}}\, dx\right)^{\frac{1}{p}} + \left(R \fint_{B(y,R)} |g|^p\, dx\right)^{\frac{1}{p}}$$

for $R < R_o$, where a constant C is independent of u and a constant R_o depends on $\|u\|_2$.

The proof is omitted here and can be found in the monograph of Giaquinta (see [**GI**]). We only point out here that the proof requires a stronger version of the reverse Hölder inequality (see Theorem 4.5), in the case where $F(t) = |t|^q$. Namely, the constant of the inequality (4.5) is replaced by a sum of maximal functions of $|f|^q$ and $|g|^q$.

CHAPTER 5

HARMONIC MEASURE

In this chapter we derive an integral representation of a solution of the Dirichlet problem involving a continuous boundary data and a Borel measure. This measure, called a harmonic measure, can be defined for elliptic operators on any bounded domain with the coefficients a^{ij} in $L^\infty(Q)$. The main purpose of this chapter is to show that if the coeficients a^{ij} satisfy a Dini condition then the solvability of the Dirichlet problem with L^2-boundary data allows us to show that the harmonic measure and the Lebesgue surface measure are mutually absolutely continuous. A counter example, due to Modica and Mortola, shows that the continuity of a^{ij} on $\bar{Q}$ is not sufficient for the absolute continuity of the harmonic measure with respect to the surface Lebesgue measure. We shall return to this question in Chapter 10.

5.1. The Dirichlet problem with continuous boundary data.

Throughout this section we assume that $Q \subset R_n$ is a bounded domain. On Q we consider the elliptic operator, given by (1.1), whose coefficients satisfy the condition (A_1). Moreover, we assume that $d^i \in L^n(Q)$, $b^i \in L^r(Q)$, $i = 1,...,n)$, and $c \in L^{\frac{r}{2}}$ for some $r > n$.

We commence with some known facts from the theory of elliptic operators.

A function $u \in W^{1,2}(Q)$ is said to be non negative on ∂Q (in the sense of $W^{1,2}(Q)$) if there exists a sequence u_m of functions in $C^1(\bar{Q})$ such that (i) $u_m \geqq 0$ on ∂Q, (ii) $u_m \to u$ in $W^{1,2}(Q)$.

Let u and v be in $W^{1,2}(Q)$. We say that $u \leqq v$ on ∂Q in $W^{1,2}(Q)$, if $v - u \geqq 0$ on ∂Q in $W^{1,2}(Q)$. In particular, v may be constant, which leads to the definition

$$\sup_{\partial Q} u = \inf\{M \in R; \quad u \leqq M \quad \text{on} \quad \partial Q \quad \text{in} \quad W^{1,2}(Q)\}.$$

Lemma 5.1. *(Caccioppoli inequality) Let $u \in W^{1,2}_{\text{loc}}(Q)$ be solution of the equation (1.1) (with $f \equiv 0$). Then for every ball $\overline{B(y,R)} \subset Q$ and all $0 < s < R$ we have*

$$\int_{B(y,s)} |Du(x)|^2\,dx \leqq \frac{C}{(R-s)^2} \int_{B(y,R)} u(x)^2\,dx, \tag{5.1}$$

where $C > 0$ is a constant depending on the norms of the coefficients of L.

PROOF. Let ψ be Lipschitz function on R_n, such that $\psi = 1$ on $B(y,s)$ and $\psi = 0$ on $R_n - B(y,R)$. Taking $v = u\psi^2$ as a test function and using the Hölder and Sobolev inequalities one can derive the estimate

$$\int_Q \psi^2 |Du|^2\,dx \leqq K \int_Q (\psi^2 + |D\psi|^2) u^2\,dx$$

for some $K > 0$. This inequality implies (5.1).

For the proofs of the next three theorems we refer to [**ST1**].

Theorem 5.1. *(Maximum principle) Suppose that*

$$c(x) - D_i b^i(x) \geqq c_\circ > 0 \quad on\ Q \tag{5.2}$$

in the sense of distribution. If $u \in W^{1,2}(Q)$ is a solution of the equation $Lu = 0$, then

$$\min\{\inf_{\partial Q} u, 0\} \leqq u(x) \leqq \max\{\sup_{\partial Q} u, 0\}.$$

Theorem 5.2. *(Hölder continuity) Let $u \in W^{1,2}_{\text{loc}}(Q)$ be a solution of $Lu = 0$. Then u satisfies a Hölder condition on each compact subset $G \subset Q$, that is, there exist constants $C > 0$ and $0 < \lambda < 1$, depending on G, such that*

$$|u(x) - u(y)| \leqq C\|u\|_2 |x - y|^\lambda.$$

Theorem 5.3. *(Harnack inequality) Let $u \in W^{1,2}_{\text{loc}}(Q)$ be a solution of $Lu = 0$. Then for each compact $G \subset Q$ there exists a positive constant K such that*

$$\max_G u \leqq K \min_G u.$$

Let us assume that (5.2) holds. It follows from Theorems 1.1 and 5.1 that if φ is the trace on ∂Q of the function $\varphi_1 \in W^{1,2}(Q)$, then there exists a unique solution $u \in W^{1,2}(Q)$ of the problem

$$Lu = 0 \text{ in } Q,\ u(x) = \varphi(x) \text{ on } \partial Q, \tag{5.3}$$

which is locally Hölder continuous, by Theorem 5.2. The boundary values are attained in the sense that $u - \varphi_1 \in \overset{\circ}{W}{}^{1,2}(Q)$. Obviously, the solution u is the same if instead of φ_1 we consider a new function φ_2 , such that $\varphi_1 - \varphi_2 \in \overset{\circ}{W}{}^{1,2}(Q)$. We denote by $T^{1,2}$ the quotient space $W^{1,2}(Q)/\overset{\circ}{W}{}^{1,2}(Q)$ equipped with the norm

$$\|\varphi\|_{T^{1,2}} = \inf_{\varphi_1 - \varphi \in \overset{\circ}{W}{}^{1,2}(Q)} \|\varphi_1\|_{W^{1,2}}.$$

This observation allows us to define a continuous linear mapping B of $T^{1,2}$ into $W^{1,2}$. By Theorem 5.1, if φ is bounded on ∂Q in the sense of $W^{1,2}(Q)$, then

$$\inf_{\partial Q} \varphi \leqq \inf_Q B\varphi \leqq \sup_Q B\varphi \leqq \sup_{\partial Q} \varphi. \tag{5.4}$$

By virtue of Lemma 5.1 and the inequality (5.4) we obtain

$$\|B\varphi\|_\# \leqq C \max_{\partial Q} |\varphi|, \tag{5.5}$$

where

$$\|g\|_{\#} = \sup_{\bar{Q}' \subset Q} \delta \left(\int_{\bar{Q}'} |Dg(x)|^2 \, dx \right)^{\frac{1}{2}} + \sup_Q |g|,$$

with δ denoting the distance between $\bar{Q}'$ and ∂Q.

From this discussion we conclude that $B\varphi$ is a linear mapping of the subset $B^{1,2}$ of functions in $T^{1,2}$ which are bounded on ∂Q (in the sense of $W^{1,2}(Q)$), into the space of functions u such that $\|u\|_{\#} < \infty$. Since any continuous function φ on ∂Q can be approximated by a uniformly convergent sequence of polynomials on ∂Q, we conclude that the set $B^{1,2}$ is dense in the space $C(\partial Q)$ equipped with the norm $\max_{\partial Q} |\varphi|$. Therefore, the linear mapping $u = B\varphi$, restricted to $C(\partial Q) \cap T^{1,2}$, can be extended to $C(\partial Q)$. The mapping so obtained is still denoted by $u = B\varphi$. This mapping assigns to every function $\varphi \in C(\partial Q)$ a unique locally Hölder continuous function u such that $\|u\|_{\#} < \infty$. It is obvious that u is a solution in $W^{1,2}_{\text{loc}}(Q)$ of the problem (5.3). Consequently we have proved the following result.

Theorem 5.4. *There exists a mapping $B\varphi$, which to any function $\varphi \in C(\partial Q)$ assigns a solution $u \in W^{1,2}_{\text{loc}}(Q)$ of the problem (5.3) (which is locally Hölder continuous on Q) in such a way that if φ is the trace of function in $W^{1,2}(Q)$, then $u = B\varphi$ coincides with solution in $W^{1,2}(Q)$ of the same problem.*

Now the question arises whether the solution $B\varphi$ approaches boundary values φ. Any point $x_\circ \in \partial Q$, where $\lim_{x \to x_\circ} B\varphi(x) = \varphi(x_\circ)$ for every $\varphi \in C(\partial Q)$ is called a regular point for the operator L. It is known that if ∂Q is of class C^2, then every point of ∂Q is regular. This means that a solution obtained by above procedure belongs to $C(\bar{Q}) \cap W^{1,2}_{\text{loc}}(Q)$. For criterions of regularity of boundary points we refer to the papers [**LSW**] and [**ST1**].

We point out that if a^{ij} satisfy the assumption (A_2) of Section 1.2 and ∂Q is of class C^2, then by Theorem 3.6 there exists a solution $u \in W^{1,2}_{\text{loc}}(Q)$ of the problem (5.3) with $\varphi \in C(\partial Q)$. It is clear that this solution coincides with the solution obtained above. Moreover, by the energy estimate (see Theorem 3.4) we have

$$\int_Q |Du(x)|^2 r(x) \, dx < \infty,$$

and this implies that

$$\sup_{\bar{Q}' \subset Q, \text{dist}(Q', \partial Q) = \delta} \delta \int_{\bar{Q}'} |Du(x)|^2 \, dx < \infty.$$

This estimate of Du is better than that given by (5.5). However, if ∂Q is of class C^2, then this estimate can be made more precise. Namely, we have

Theorem 5.5. *If ∂Q is of class C^2, then for a bounded solution $u \in W^{1,2}_{\text{loc}}(Q)$ of (5.3) we have*

$$\int_Q |Du(x)|^2 r(x)^{1+s} \, dx < \infty.$$

for every $s > 0$.

PROOF. We only consider the case $0 < s < 1$. Let us now observe that

$$v(x) = \begin{cases} u(x)(\rho(x) - \delta)^{1+s} & \text{on } Q_\delta \\ 0 & \text{on } Q - Q_\delta \end{cases}$$

is in $\overset{\circ}{W}{}^{1,2}(Q)$. Consequently, taking v as a test function we obtain

$$\int_Q a^{ij} D_i u D_j u(\rho - \delta)^{1+s}\, dx + (1+s) \int_Q a^{ij} D_i u u(\rho - \delta)^s D_j \rho\, dx$$
$$+ \int_Q (b^i + d^i) D_i u u(\rho - \delta)^{1+s}\, dx + (1+s) \int_Q b^i D_i \rho(\rho - \delta)^s u^2\, dx$$
$$+ \int_Q c u^2 (\rho - \delta)^{1+s}\, dx = \int_Q f u(\rho - \delta)^{1+s}\, dx.$$

Applying the ellipticity condition, the Hölder and Sobolev inequality and the fact that $0 < s < 1$ we easily arrive at the estimate

$$\int_{Q_\delta} |Du(x)|^2 (\rho(x) - \delta)^{1+s}\, dx \leqq C_1 \int_Q |f|\, dx + C_2$$

with a constants $C_1 > 0$ and $C_2 > 0$ independent of δ. Letting $\delta \to$) the result easily follows.

5.2. Harmonic measure and the Lebesgue surface measure.

Let L be the elliptic operator given by (1.1) and stisfying the assumptions of the previous section. Moreover we assume that (5.2) holds. By Theorem 5.4 the Dirichlet problem (5.3) is solvable for any continuous function $\varphi \in C(\partial Q)$. The maximum principle (Theorem 5.1) implies that for each fixed $x \in Q$ the mapping $\varphi \to u(x)$ is a positive linear functional on $C(\partial Q)$. Therefore there exists a Borel measure W_L^x on ∂Q such that

$$u(x) = \int_{\partial Q} \varphi(y)\, dW_L^x(y).$$

The aim of this section is to show that if the coefficients a^{ij} satisfy the Dini condition (see (A_2), Section 2.1), then the harmonic measure W_L^x and the Lebesgue surface measure are mutually absolutely continuous. To prove this we need the following result on the local boundedness of solutions in $W^{1,2}_{\text{loc}}(Q)$ of $Lu = 0$.

Theorem 5.6. *Let $u \in W^{1,2}_{\text{loc}}(Q)$ be a solution of $Lu = 0$ and suppose that $B(x, R) \subset Q$, Then there exists a constant $K > 0$ depending on the norms of coefficients such that*

$$|u(y)| \leqq K \left(R^{-n} \int_{B(x,R)} u(z)^2\, dz \right)^{\frac{1}{2}}$$

for all $y \in B(x, \frac{R}{2})$.

For the proof we refer to [LU], [ST1] or [SE].

Theorem 5.7. *Suppose that ∂Q is of class C^2 and that the coefficients a^{ij}, i,j=1,...,n, satisfy (A_2). Then the harmonic measure W_L^x and the surface Lebesgue measure are mutually absolutely continuous.*

PROOF. Let u be a solution of the Dirichlet problem (5.3) in $\widetilde{W}^{1,2}(Q)$ with a boundary data $\varphi \in L^2(\partial Q)$. Let us fix $x \in Q$ with $B(x,r) \subset Q$. By virtue of Theorem 5.6,

$$|u(y)| \leqq K\left(r^{-n}\int_{B(x,r)} u(z)^2\,dz\right)^{\frac{1}{2}}$$

for $y \in B(x, \frac{r}{2})$. Now by the energy estimate (3.30) (see Theorem 3.5) we obtain

$$|u(y)| \leqq KC\left(\int_{\partial Q} \varphi(x)^2\,dS_x\right)^{\frac{1}{2}}$$

and the absolute continuity of the L-harmonic measure W_L^x with the respect to the Lebesgue surface measure follows the Riesz representation theorem of a linear functional on $L^2(\partial Q)$. Consequently

$$W_L^x(E) = \int_E k(x,y)dS_y$$

for every Borel measure $E \subset \partial Q$, where $k(x,\cdot) \in L^2(\partial Q)$ denotes the density of W_L^x. Suppose that there exists a Borel set $E \subset \partial Q$ such that

$$W_L^{x_o}(E) = 0 \quad \text{and} \quad S(E) > 0$$

for certain $x_o \in Q$, with S denoting the surface Lebesgue measure. By Harnack's inequality (see Theorem 5.2) all measures $\{W_L^x\}$ are mutually absolutely continuous and consequently $W_L^x(E) = 0$ for all $x \in Q$. Consider now the Dirichlet problem (5.3) with the boundary data $\varphi(x) = \chi_E(x)$, where χ_E denotes the characteristic function of E. The solution of this problem is given by

$$u(x) = \int_{\partial Q} k(x,y)\chi_E(y)\,dS_y = \int_E dW_L^x(y) = W_L^x(E) = 0$$

for all $x \in Q$. Now the L^2-convergence $u(x_\delta) \to \chi_E$ gives a contradiction, since $S(E) > 0$.

5.3 Counter example of Modica and Mortola.

Modica and Mortola (see [MM]) constructed an example of an elliptic operator with a singular harmonic measure. In their example this operator is given by

$$\tilde{L}u = D_x^2 u + D_y(\alpha(x,y)D_y u) \tag{5.4}$$

on an open bounded connected set $Q \subset R_2$ with the boundary of class C^∞ such that

$$\partial Q \supset T = \{(x,y) \in R_2;\quad -a \leqq x \leqq a,\quad y = 0\}$$

for some $a > 0$ and $\alpha \in C(R_2) \cap C^\infty(R_2 - \{y = 0\})$.

The construction of a singular measure is based on the following result from the theory of the Fourier - Stieltjes series.

Let $\{n_k\}$ be sequence of positive integers satisfying a condition $\frac{n_{k+1}}{n_k} \geqq q > 1$ and let $\{\alpha_k\}$ be a sequence of constants such that $-1 \leqq \alpha_k \leqq 1$, $\alpha_k \neq 0$ for all $k > 1$. Let us now define

$$p_k(t) = \prod_{j=1}^{k} (1 + \alpha_j \cos n_j t),$$

where $t \in R$.

Theorem 5.8. *If $\frac{n_{k+1}}{n_k} \geqq q > 3$, and $\sum_{j=1}^\infty \alpha_j^2 = \infty$, then the sequence p_k converges weakly to a singular measure ν, as $k \to \infty$.*

For the proof we refer to the monograph [**ZY**] (pp.208-209).

We now return to the construction of a singular harmonic measure. Following Modica and Mortola [MM] we define

$$\alpha(x,y) = \begin{cases} \Phi_1(x) & \text{if} \quad |y| \geqq k_1^{-1}, \\ \psi(k_{n+1}y)\Phi_{n+1}(x) + (1 - \psi(k_{n+1}y))\Phi_n(x) & \text{if} \quad k_{n+1}^{-1} \leqq |y| < k_n^{-1}, \\ & \qquad n = 1, 2, ... \\ 1 & \text{if} \quad y = 0, \end{cases}$$

where $\Phi_n(x) = 1 + \frac{1}{2\sqrt{n}} \cos h_n x$, $\{h_n\}$, $\{k_n\}$ are suitable sequence of positive integers such that $h_{n+1} \geqq 4h_n$ and $k_{n+1} \geqq 2k_n$ for every n and ψ is a function in $C_0^\infty(R)$ satisfying the conditions: $\psi(t) = \psi(-t)$, and $0 \leqq \psi(t) \leqq 1$ for all $t \in R$, $\psi(t) = 1$ for $|t| \leqq 1$ and $\psi(t) = 0$ for $|t| \geqq 2$. The sequences $\{h_n\}$ and $\{k_n\}$ will be constructed by a compactness argument. Note that $\alpha \in C(R^2) \cap C^\infty(R_2 - \{y = 0\})$ and $\frac{1}{2} \leqq \alpha(x,y) \leqq \frac{3}{2}$ for all $(x,y) \in R_2$.

Let $\{\alpha_n(x,y)\}$ be a sequence of functions in $C^\infty(R^2)$ given by

$$\alpha_n(x,y) = \begin{cases} \alpha(x,y) & \text{for} \quad |y| \geqq k_n^{-1} \\ \Phi_n(x) & \text{for} \quad |y| < k_n^{-1} \quad (n = 1, 2, ...), \end{cases}$$

which converges uniformly on R_2 to α. It is clear that $\frac{1}{2} \leqq \alpha_n(x,y) \leqq \frac{3}{2}$ for all R_2. We define the operators $\tilde{L}_n$ by

$$\tilde{L}_n = D_x^2 u + D_y(\alpha_n(x,y) D_y u).$$

We now associate with $\tilde{L}$ and $\tilde{L}_n$ harmonic measures W_P and W_P^n, respectively, evaluated at $P \in Q$. We always assume that

$$Q \supset B^+(0,c) = \{(x,y); \quad x^2 + y^2 < c^2, \quad y > 0\}$$

for some $c > a$ and $P \in Q - B^+(0,c)$.

Lemma 5.2. *The sequence $\{W_P^n\}$ converges weakly in the sense in the sense of measures to W_P, that is,*

$$\lim_{n\to\infty}\int_{\partial Q}\chi\, dW_P^n = \int_{\partial Q}\chi\, dW_P \tag{5.5}$$

for every $\chi \in C(\partial Q)$.

PROOF. Since $\int_{\partial Q} dW_p^n = 1$ for all n, it suffices to prove (5.5) for $\chi \in C^\infty(\partial Q)$. Let $\chi \in C^\infty(\bar{Q})$ and let u be a unique solution in $W^{1,2}_{\mathrm{loc}}(Q) \cap C(\bar{Q})$ to the Dirichlet problem

$$\tilde{L}u = 0 \quad \text{in } Q, \quad u = \chi \quad \text{on} \quad \partial Q.$$

Similarly, we define u_n with $\tilde{L}$ replaced by $\tilde{L}_n$. We have to prove that $\lim_{n\to\infty} u_n(P) = u(P)$. The operators $\tilde{L}$ and $\tilde{L}_n$ induce the isomorphisms E and E_n from $\overset{\circ}{W}{}^{1,2}(Q)$ onto $W^{-1,2}(Q)$ (see Theorems 1.1 and 5.1). Since

$$\frac{1}{2} \leqq \alpha_n(x,y) \leqq \frac{3}{2} \quad n = 1, 2, \dots$$

for all $(x,y) \in R_2$ and $\alpha_n \to \alpha$ in $L^2(Q)$, there exist positive constants d_1 and d_2 such that

$$d_1 \leqq \|E_n\| \leqq d_2 \quad \text{and} \quad d_2^{-1} \leqq \|E_n^{-1}\| \leqq d_1^{-1}$$

for all n. It is clear that $\tilde{L}_n v \to \tilde{L}v$ and $E_n^{-1}w \to E^{-1}w$ as $n \to \infty$ in $W^{-1,2}(Q)$ and $\overset{\circ}{W}{}^{1,2}(Q)$, respectively, for every $v \in W^{1,2}(Q)$ and $w \in W^{-1,2}(Q)$. Since

$$u_n = \chi + E_n^{-1}(-\tilde{L}_n\chi) \quad \text{for all } n,$$

and

$$u = \chi + E^{-1}(-\tilde{L}\chi),$$

we have

$$\|u_n - u\|_{W^{1,2}(Q)} \leqq \|E_n^{-1}\|\|\tilde{L}_n\chi - \tilde{L}\chi\|_{W^{-1,2}} + \|E_n^{-1}L\chi - E^{-1}L\chi\|_{\overset{\circ}{W}{}^{1,2}}$$

and consequently $\lim_{n\to\infty} u_n = u$ in $W^{1,2}(Q)$. Applying Theorem 5.2 the result follows.

Since for any Borel set $E \subset \partial Q$, $W_P(E)$ as a function of P is a positive solution in $W^{1,2}_{loc}(Q)$ of $\tilde{L}u = 0$, it follows from Harnack's inequality (see Theorem 5.3) that for any P and R in Q the harmonic measures W_P and W_R are mutually absolutely continuous.

We also recall the following result from the theory of Sobolev spaces. The space $W^{1,p}(Q)$, with $p > n$, is continuously imbedded in the Hölder space $C^\alpha(\bar{Q})$ with $\alpha < \frac{p-n}{p}$.

To procced further we denote by $\{\beta_n\}$ a sequence of functions in $C^\infty(R_2)$ such that $\beta_n(x,y) = \beta_n(x,-y)$ and $\frac{1}{2} \leqq \beta_n(x,y) \leqq \frac{3}{2}$ on R_2 for all n. Let

$$M_n = D_x^2 + D_y(\beta_n(x,y)D_y)$$

and denote by G_n the Green function for M_n on Q with a pole at P.

In the proof of Lemma 5.3 below, we frequently use some properties of the Green function and the Green operator for elliptic operators. For all relevant properties of the Green function and the Green operator we refer to the paper [**LSW**].

Lemma 5.3. *(i) If* $\lim_{n\to\infty}\beta_n = \beta$ *in* $L^2(Q)$ *and there exists constant* $C_1 > 0$ *such that*

$$|D_x\beta_n(x,y)| \leqq C_1 \quad \text{on } Q$$

for all n*, then*

$$\lim_{n\to\infty}\max_{|x|\leqq a}|\beta_n(x,0)D_yG_n(x,0) - \beta(x,0)D_yG(x,0)| = 0,$$

where G *is the Green function for the operator*

$$M = D_x^2 + D_y(\beta(x,y)D_y).$$

(ii) If $\lim_{n\to\infty}\beta_n = \beta$ *weakly in* $L^2(Q)$ *and there exists a constant* $C_1 > 0$ *such that*

$$|D_y\beta(x,y)| \leqq C_1 \quad \text{on} \quad Q$$

for all n*, then*

$$\lim_{n\to\infty}\max_{|x|\leqq a}|D_yG_n(x,0) - D_yG(x,0)| = 0.$$

PROOF. To establish (i) we show that there exist $\epsilon > 0$ and $C_2 > 0$ such that

$$\|\beta_n D_y G_n\|_{W^{1,2+\epsilon}(B_a^+)} \leqq C_2 \tag{5.6}$$

for all n. We extend G_n into $B(0,c)$ by setting $G_n(x,y) = G_n(x,-y)$ for all $(x,y) \in Q$. We now observe that

$$G_n \in W^{1,2}(B(0,c)) \quad \text{and} \quad M_nG_n = 0 \quad \text{on} \quad B(0,c)$$

for all n. Since $P \notin \overline{B(0,c)}$, there exists a constant $C_3 > 0$ such that

$$\|G_n\|_{W^{1,2}(B(0,c))} \leqq C_3$$

for all n (see **[LSW]** , p.69). It follows from Theorem 4.7 that there exist constants $\epsilon > 0$ and $C_4 > 0$ such that

$$\|G_n\|_{W^{1,2+\epsilon}(B(0,a))} \leqq C_4.$$

Now, differentiating the equation $M_nG_n = 0$ we obtain that

$$M_n(D_xG_n) = -D_y(D_x\beta_nD_yG_n)$$

and

$$M_n(D_yG_n) = -D_y(D_y\beta_nD_yG_n).$$

According to Theorem 4.7

$$\|D_xG_n\|_{W^{1,2+\epsilon}(B(0,a))} \leqq C_4\Big(\|D_xG_n\|_{L^2(B(0,c))}$$
$$+ \|D_x\beta_n\|_{L^\infty(B(0,c))}\|D_yG_n\|_{L^{2+\epsilon}(B(0,c))}\Big) \leqq C(C_3 + C_1C_4).$$

and

$$\|D_y G_n\|_{W^{1,2+\epsilon}(B(0,a))} \leqq C(C_3 + C_1 C_4).$$

From the last two inequalities we deduce that

$$\begin{aligned}
&\|\beta_n D_y G_n\|_{W^{1,2+\epsilon}(B(0,a))} \leqq \|\beta_n\|_{L^\infty(B(0,a))}\|D_y G_n\|_{L^{2+\epsilon}(B(0,a))} \\
&+ \|D_x(\beta_n D_y G_n)\|_{L^{2+\epsilon}(B(0,a))} + \|D_y(\beta_n D_y G_n)\|_{L^{2+\epsilon}(B(0,a))} \\
&\leqq \frac{3}{2} C(C_3 + C_1 C_4) + \|D_x \beta_n\|_{L^\infty(B(0,a))}\|D_y G_n\|_{L^{2+\epsilon}(B(0,a))} \\
&+ \frac{3}{2}\|D^2_{xy} G_n\|_{L^{2+\epsilon}(B(0,a))} + \|D^2_x G_n\|_{L^{2+\epsilon}(B(0,a))} \leqq Const.
\end{aligned}$$

This estimate and the Sobolev imbedding theorem, mentioned in the paragraph preceding this lemma, imply that the sequence $\{\beta_n D_y G_n\}$ is bounded in $C^{0,\delta}(\bar{B}^+_a)$ with $\delta = \frac{\epsilon}{2+\epsilon}$. Hence, by Ascoli's theorem we can select a subsequence of $\{\beta_n D_y G_n\}$ converging on $\bar{B}^+_a$. Therefore to prove (i) it suffices to show that $\lim_{n\to\infty} D_y G_n = D_y G$ weakly in $L^q(Q)$ for all $q \in (1,2)$.

Let $G^{(p)}_n: \quad W^{-1,p}(Q) \to C(\bar{Q})$, $2 < p < \infty$, denote the Green operator of M_n, that is

$$M_n(G^{(p)}_n u) = u \quad \text{on } Q \quad \text{and} \quad G^{(p)}_n = 0 \quad \text{on } \partial Q$$

for all $u \in W^{-1,p}(Q)$ and let $G^{(p)*}_n$ be its adjoint operator. It is known that $G_n = G^{(p)*}_n(\delta_P)$, where δ_P is the Dirac measure concentrated at P. Hence $G_n \in \overset{\circ}{W}{}^{1,q}(Q)$ for every $q \in (1,2)$ and all n. Define by $\tilde{G}_n$ the mapping of $\bigcup_{p>2} W^{-1,p}(Q)$ onto $C(\bar{Q})$ defined by $G^{(p)}_n$. In the obvious way we define $\tilde{G}$. The weak convergence of $\{G_n\}$ to G in $\overset{\circ}{W}{}^{1,q}(Q)$ is equivalent to $\lim_{n\to\infty} \tilde{G}_n u(P) = \tilde{G}u(P)$ for all $u \in W^{-1,p}(Q)$, where $p = \frac{q}{q-1}$. By Theorem 4.7, there exists $\epsilon > 0$ (depending on p and Q) such that $\tilde{G}u \in \overset{\circ}{W}{}^{1,2+\epsilon}(Q)$ for every $u \in W^{-1,p}(Q)$. Hence the definition of $G^{2+\epsilon}_n$ implies that

$$G^{(2+\epsilon)}_n(M_n \tilde{G}u) = \tilde{G}u$$

for all $u \in W^{-1,p}(Q)$ and all n. Therefore, for every $u \in W^{-1,p}(Q)$ and all n we have

$$\begin{aligned}
|(\tilde{G}_n u)(P) - (\tilde{G}u)(P)| &\leqq \|\tilde{G}_n u - \tilde{G}u\|_{C(\bar{Q})} \\
&= \|G^{(2+\epsilon)}_n(M_n \tilde{G}u - M\tilde{G}u)\|_{C(\bar{Q})} \\
&\leqq \|G^{(2+\epsilon)}_n\| \|M_n(\tilde{G})u - M(\tilde{G}u)\|_{W^{-1,2+\epsilon}(Q)},
\end{aligned}$$

where $\|G^{(2+\epsilon)}_n\|$ denotes the norm of $G^{(2+\epsilon)}_n$ as an operator from $W^{-1,2+\epsilon}(Q)$ into $C(\bar{Q})$. Since these norms are bounded (see [LSW], p.60), it remains to prove that

$$\lim_{n\to\infty} \|M_n u - Mu\|_{W^{-1,p}(Q)} = 0 \tag{5.7}$$

for all $u \in W^{1,p}(Q)$ and $p \in (2,\infty)$. The equicontinuity of $M_n: \quad W^{1,p}(Q) \to W^{-1,p}(Q)$ allows us assume that $u \in C^\infty(\bar{Q})$. Since the relation (5.7) automatically holds if $\lim_{n\to\infty} \beta_n = \beta$ in $L^2(Q)$, the assertion follows.

To prove (ii) we observe that, since β_n converges to β weakly in $L^2(Q)$ and $|D_y\beta|$ is bounded, $M_n u$ converges to Mu in the sense of distributions. On the other hand $M_n u$ is bounded in $L^p(Q)$ and by the Sobolev imbedding theorem $M_n u$ converges to $M_n u$ in $W^{-1,2}(Q)$, as $n \to \infty$, and this completes the proof.

We are now in a position to construct an elliptic operator $\tilde{L}$ with a singular $\tilde{L}$-harmonic measure.

Theorem 5.8. *Let Q be an open bounded connected subset of R_2 with the boundary ∂Q of class C^∞ such that $\partial Q \supset T = \{(x,y) \in R_2;\ -a \leqq x \leqq a,\ y = 0\}$. Then there exists a function $\alpha \in C^\infty(Q) \cap C(\bar{Q})$, with $0 < \lambda \leqq \alpha(x,y) \leqq \Lambda$ in Q for some positive constants λ and Λ, such that, if*

$$\tilde{L} = D_x^2 + D_y(\alpha(x,y)D_y),$$

then for each P, the $\tilde{L}$-harmonic measure W_P is not absolutely continuous with respect to the Lebesgue measure on T.

PROOF. First we observe that by Green's theorem we have

$$\int_{\partial Q} \chi\, dW_P = \int_{-c}^{c} \chi(x,0)\Phi_n(x) D_y G_n(x,0)\, dx$$

for every n and $\chi \in C(\partial Q)$ with compact support in T, where $G_n \in C^\infty(Q - \{P\})$ denotes the Green function in Q for the operator $\tilde{L}_n$ with the pole at P.
Let

$$K = \min_{|x| \leqq a} |D_y G_1(x,0)|.$$

It follows from a classical boundary point lemma (see [**LSW**], p.65) that $K > 0$. We now show that there exist sequences $\{h_n\}$ and $\{k_n\}$ of positive integers such that

$$h_{n+1} \geqq 4h_n, \quad k_{n+1} \geqq 2k_n \quad \text{for all} \quad n, \tag{5.8}$$

and

$$\max_{|x| \leqq a} |D_y G_{n+1}(x,0) - \Phi_n(x) D_y G_n(x,0)| \leqq K 4^{-n-1} \tag{5.9}$$

for all n. Suppose that $h_1 = 1, h_2, ..., h_n$, $k_1 = 1, k_2, ..., k_n$ have already been chosen and let us construct h_{n+1} and k_{n+1}. Let k be a positive integer such that $k \geqq 2k_n$ and put

$$\beta_k(x,y) = \begin{cases} \alpha_n(x,y) & \text{for} \quad |y| \geqq k_n^{-1}, \\ \psi(ky) + (1 - \psi(ky))\Phi_n(x) & \text{for} \quad |y| < k_n^{-1}. \end{cases}$$

It is easy to see that

$$\beta_k \in C^\infty(R_2), \quad \beta_k(x,y) = \beta_k(x,-y), \quad \frac{1}{2} \leqq \beta_k(x,y) \leqq \frac{3}{2}$$

for all $k \geqq 2k_n$ and $(x,y) \in R_2$. Moreover, $\lim_{k\to\infty} \beta_k = \alpha_n$ in $L^2(Q)$ and there exists a constant $C_1 > 0$ such that

$$|D_x\beta_k(x,y)| \leqq C_1 \quad \text{for all} \quad k \geqq 2k_n \quad \text{and} \quad (x,y) \in Q.$$

By virtue of Lemma 5.3 there exists an integer $\bar{k}$ such that

$$\bar{k} \geqq 2k_n \tag{5.10}$$

and

$$\begin{aligned} &\max_{|x|\leqq a} |\beta_{\bar{k}}(x,0)D_y\bar{G}_k(x,0) - \alpha_n(x,0)D_yG_n(x,y)| \\ &= \max_{|x|\leqq a} |D_y\bar{G}_k(x,0) - \Phi_n(x)D_yG_n(x,0)| \leqq K4^{-n-2}, \end{aligned} \tag{5.11}$$

where $\bar{G}_k$ denotes the Green function on Q for the operator $D_x^2 + D_y(\beta_{\bar{k}}D_y)$ with the pole at P. Let us set $k_{n+1} = \bar{k}$ and for a positive integer h we consider the function $\tilde{\beta}_h(x,y)$ defined by

$$\tilde{\beta}_h(x,y) = \begin{cases} \alpha_n(x,y) & \text{for } |y| \geqq k_n^{-1}, \\ \psi(k_{n+1}y)(1 + \frac{1}{2\sqrt{n+1}}\cos(hx)) + (1 - \psi(k_{n+1}y))\Phi_n(x) & \\ & \text{for} \quad |y| < k_n^{-1}. \end{cases}$$

It is clear that

$$\tilde{\beta}_h \in C^\infty(R_2), \quad \tilde{\beta}_h(x,y) = \tilde{\beta}_h(x,-y), \quad \frac{1}{2} \leqq \tilde{\beta}_h(x,y) \leqq \frac{3}{2}$$

for all positive integers h and $(x,y) \in R_2$. Moreover, $\lim_{h\to\infty} \tilde{\beta}_h = \tilde{\beta}_{\bar{k}}$ weakly in $L^2(Q)$ and there exists a positive constant C_1 such that

$$|D_y\tilde{\beta}_h(x,y)| \leqq C_1$$

for all positive integers h and $(x,y) \in Q$. It follows from Lemma 5.3 that there exists a positive integer $\bar{h}$ such that

$$\bar{h} \geqq 4k_n \tag{5.12}$$

and

$$\max_{|x|\leqq a} |D_y\tilde{G}_{\bar{h}}(x,0) - D_yG_{\bar{h}}(x,y)| \leqq K4^{-n-2}, \tag{5.13}$$

where $\tilde{G}_{\bar{h}}$ denotes the Green function for the operator $D_x^2 + D_y(\tilde{\beta}_{\bar{h}}D_y)$ with the pole at P. We now define $h_{n+1} = \bar{h}$. Then $\tilde{\beta}_{\bar{h}} = \alpha_{n+1}$ and $\tilde{G}_{\bar{h}} = G_{n+1}$, and consequently (5.10), (5.11), (5.12) and (5.13) imply (5.8) and (5.9). To complete the proof we set

$$w_n(x) = D_yG_{n+1}(x,0)\left[\prod_{i=1}^{n} \Phi_i(x)\right]^{-1}$$

for $|x| \leqq a$ and $n = 1, 2, \ldots$. It is easy to check that

$$w_n(x) = D_y G_1(x,0) + \sum_{j=1}^{n} \Big(D_y G_{j+1}(x,0) - \Phi_j(x) D_y G_j(x,0) \Big) \left[\prod_{i=1}^{j} \Phi_i(x) \right]^{-1}.$$

Therefore the inequality (5.9) and the fact that $\Phi_n(x) \geqq \frac{1}{2}$ for all n imply that the sequence $\{w_n\}$ converges uniformly on $[-a, a]$ to a function w and moreover

$$\max_{|x| \leqq a} |w_n(x) - w(x)| \leqq K 2^{-n-2} \tag{5.14}$$

for all n and

$$\frac{3}{4} K \leqq w(x) \tag{5.15}$$

for all $x \in [-a, a]$. Let $\chi \in C(\partial Q)$ with compact support in T. Then

$$\begin{aligned} \int_{-a}^{a} \chi(x) \Phi_n(x) D_y G_n(x,0)\, dx = & \int_{-a}^{a} \chi(x)(w_{n_1}(x) - w(x)) \prod_{i=1}^{n} \Phi_i(x)\, dx \\ & + \int_{-a}^{a} \chi(x) w(x) \prod_{i=1}^{n} \Phi_i(x)\, dx. \end{aligned}$$

Since $\Phi_i \leqq \frac{3}{2}$ for all i, it follows from (5.14) that the first integral on the right side tends to 0. According to Theorem 5.8 the second integral converges to

$$\int_{-a}^{a} \chi(x) w(x)\, d\nu,$$

where ν is a singular measure with respect to the Lebesgue measure on $[-a, a]$.

CHAPTER 6

EXCEPTIONAL SETS ON THE BOUNDARY

In this chapter we discuss the question of the uniqueness of the Dirichlet problem. Namely, we suppose that a solution assumes boundary values on ∂Q except on a set $E \subset \partial Q$. We prove that for $\frac{n}{n-1} \leqq p < \infty$, the space $L^p(Q)$ is a class of uniqueness for the Dirichlet problem if E has a finite Hausdorff measure of order $n - q$, where $\frac{1}{p} + \frac{1}{q} = 1$. We also give an example showing that the order of the Hausdorff measure is optimal. The results of this chapter are due to Gaǰdenko (see [**GA**]).

6.1. Formulation of the problem and preliminaries.

In Chapter 3 we have constructed a solution to the Dirichlet problem in $W^{1,2}_{loc}(Q)$, where the boundary data $\varphi \in L^2(\partial Q)$ is recovered in the sense of the L^2-convergence

$$\lim_{\delta \to 0} \int_{\partial Q} [u(x_\delta) - \varphi(x)]^2 \, dS_x = 0. \tag{6.1}$$

(See Section 3.4). Since ∂Q is a compact set, it is clear that the relation (6.1) holds if and only if each point $x_\circ \in \partial Q$ has a neighborhood $U(x_\circ) \subset \partial Q$ such that

$$\lim_{\delta \to 0} \int_{U(x_\circ)} [u(x_\delta) - \varphi(x)]^2 \, dS_x = 0. \tag{6.2}$$

The main purpose of this chapter is to investigate solutions in $W^{1,2}_{\rm loc}(Q)$ of equation (1.1) (with $f \equiv 0$) assuming boundary values in the sense of (6.2) for almost all points $x_\circ \in \partial Q$.

Namely, let us denote by E the set of points of ∂Q at which the relation (6.2) does not hold. The following question arises: under what conditions on the "dimension" of the exceptional set E will the space $L^p(\partial Q)$, $1 < p < \infty$, be a uniqueness class for the Dirichlet problem.

In particular, from Theorem 3.7 follows automatically the following uniqueness result.

Let $u \in W^{1,2}_{\rm loc}(Q) \cap L^2(Q)$ be a solution of $Lu = 0$. Suppose that for every $x_\circ \in \partial Q$ there exists a neighborhood $U(x_\circ) \subset \partial Q$ such that

$$\lim_{\delta \to 0} \int_{U(x_\circ)} u(x_\delta)^2 \, dS_x = 0.$$

Then $u \equiv 0$ on Q.

In this case $E = \varnothing$. In what follows we impose a condition on E under which $L^p(\partial Q)$, $1 < p < \infty$, is a uniqueness class for solutions in $W^{1,2}_{\rm loc}(Q)$.

We begin with the following construction. Let $p \in (\frac{n}{n-1}, \infty)$ and let E be a closed subset of ∂Q, of a finite Hausdorff measure of order $n - q$, with $\frac{1}{p} + \frac{1}{q} = 1$. We denote by G_k a lattice in R_n consisting of closed cubes with edges of length $a_k = 2^{-k}$, the vertices of which have coordinates of the form $N2^{-k}$, where N is an integer. Since E is a closed set and its Hausdorff measure of order $n-q$ is less than a positive number c_o, for every sufficiently small $\epsilon > 0$ there is a finite covering of the set E by balls B_i, $i = 1, ..., m'(\epsilon)$ of radii $r_i < \epsilon$, such that $\sum_{i=1}^{m'(\epsilon)} r_i^{n-q} < c_o$. For each ball B_i there is a positive number $k(i)$ such that $2^{-(k(i)+1)} \leqq r_i < 2^{-k(i)}$. We denote by $G^{(i)}$ the set of cubes of the lattice $G_{k(i)}$ which have non-empty intersection with B_i, it is clear that $G^{(i)}$ contains no more than 3^n elements. Taking the union of all sets $G^{(i)}$ with $i = 1, ..., m'(\epsilon)$, we obtain a covering of the set E by a family of closed cubes belonging to different lattices. We exclude from this family those cubes contained entirely within other cubes of the family and as well as cubes not containing points of E. Consequently we obtain a covering of E by family of closed cubes $\{K_l\}$, $l = 1, ..., m(\epsilon)$, with $m(\epsilon) \leqq 3^n m'(\epsilon)$, which possess the following properties:

(i) $\operatorname{Int} K_l \cap \operatorname{Int} K_{l'} = \varnothing$ for $l \neq l'$,

(ii) each cube K_l belongs to some lattice G_k, $l = 1, ..., m(\epsilon)$, with $a_{k_l} = 2^{-k_l} \leqq 2\epsilon$,

(iii) $\sum_{l=1}^{m(\epsilon)} a_{k_l}^{n-q} \leqq 3^n \sum_{i=1}^{m'(\epsilon)} (2r_i)^{n-q} \leqq 3^n 2^{n-q} c_o$.

We refer to $\{K_l\}$, $l = 1, ..., m(\epsilon)$, as a covering constructed for the given ϵ. Obviously this covering fulfils conditions (i), (ii) and (iii).

We denote by λK_l the open cube with the same center as K_l but with edges of length λa_k with $\lambda > 0$.

We need the following form of a partition of unity.

Lemma 6.1. *Let ϵ be a given positive number, and let $\{K_l\}$ $(l = 1, ..., m)$ be covering of the set E constructed for ϵ , ordered by decreasing lengths of their edges: $a_{k_1} \geqq a_{k_2} \geqq ... \geqq a_{k_m}$, $m = m(\epsilon)$. There exists a family of functions $\{\xi_l\}$ $(l = 1, ..., m)$ such that $\xi \in C^\infty(R_n)$, $\operatorname{supp} \xi_l \subset 2K_l$ $(l = 1, ..., m)$ and $\sum_{l=1}^m \xi_l(x) = 1$ for $x \in \bigcup_{l=1}^m \frac{3}{2} K_l$. Moreover, there is a constant C depending only on n such that for each $l = 1, ..., m$ and for all $x \in R_n$ the inequalities*

$$|D_i \sum_{r=1}^{l} \xi_r(x)| \leqq \frac{C}{a_{k_l}} \qquad i = 1, ..., n$$

and

$$|D_{ij} \sum_{r=1}^{l} \xi_r(x)| \leqq \frac{C}{a_{k_l}^2} \qquad (i, j = 1, ..., n)$$

hold.

PROOF. We choose a function $\eta \in C^\infty(R_n)$ such that $0 \leqq \eta(x) \leqq 1$ for all $x \in R_n$, $\eta(x) = 1$ for $|x_i| \leqq \frac{3}{2}$ $(i = 1, ..., n)$ and $\eta(x) = 0$ when $|x_i| > 2$ for some i. We set

$$\eta_l(x) = \eta\left(\frac{2(x - x^l)}{a_{k_l}}\right),$$

where x^l is the centre of K_l and define the functions $\xi_r(x)$ in the following manner:

$$\xi_1 = \eta_1 \text{ and } \xi_l = \eta_l \prod_{r=1}^{l-1} (1 - \eta_r) \text{ for } l = 2, ..., m.$$

Then $\xi_l \in C^\infty(R_n)$ and $\operatorname{supp}\xi_l \subset \operatorname{supp}\eta_l \subset 2K_l$. By induction, it is easy to verify that

$$\sum_{r=1}^{l} \xi_r = 1 - \prod_{r=1}^{l}(1-\eta_r) \tag{6.3}$$

for each $l = 1, ..., m$, and therefore $\sum_{r=1}^{m} \xi_r(x) = 1$ for $x \in \bigcup_{r=1}^{m} \frac{3}{2}K_r$.
To establish the estimates for the derivatives we observe that

$$|D_i \sum_{r=1}^{l} \xi_r(x)| \leqq \sum_{r=1}^{l} |D_i\eta_r(x)|.$$

We note that not more than 2^n open cubes of the same size from the same lattice have non-empty intersection. Therefore among those cubes of the family $\{2K_l\}$ $(l = 1, ..., m)$ which contain a point $x \in R_n$, not more than 2^n of them have equal edges. Moreover

$$|D_i\eta_r(x)| \leqq \max_{R_n} |D\eta(x)| \frac{2}{a_{k_r}}, \quad a_{k_r} = a_{k_l} 2^{k_l - k_r}$$

(by assumption $k_l \geqq k_r$ for $r \leqq l$). Consequently, for every point $x \in R_n$ we have

$$\sum_{r=1}^{l} |D_i\eta_r(x)| \leqq 2 \max_{R_n} |\eta(x)| \frac{2^n}{a_{k_l}} \sum_{k=0}^{\infty} 2^{-k}.$$

Similarly

$$\begin{aligned} |D_{ij} \sum_{r=1}^{l} \xi_r(x)| &\leqq \sum_{r=1}^{l} |D_{ij}\eta_r(x)| + \Big(\sum_{r=1}^{l} |D_i\eta_r(x)|\Big)\Big(\sum_{r=1}^{l} |D_i\eta_r(x)|\Big) \\ &\leqq \max_{\{i,j=1,...,n\}} \big(\max_{R_n} |D_{ij}\eta(x)|\big) 2^{n+2} \frac{1}{a_{k_l}^2} \sum_{k=0}^{\infty} 2^{-2k} \\ &+ \Big(\max_{R_n} |D\eta(x)| \frac{2^{n+1}}{a_{k_l}} \sum_{k=0}^{\infty} 2^{-k}\Big)^2. \end{aligned}$$

6.2. Uniqueness criterion for solutions in $W^{1,2}_{\text{loc}}(Q) \cap L^p(Q)$.

Throughout this chapter we assume that L is uniformly elliptic in Q and moreover $a^{ij} \in C^1(\bar{Q})$, $b^i \in C^1(\bar{Q})$ $(i, j = 1, ..., n)$, d^i $(i = 1, ..., n)$ and $c \in L^\infty(Q)$. Under these assumptions every solution $u \in W^{1,2}_{\text{loc}}(Q)$ of the equation $Lu = 0$ belongs to $W^{2,2}_{\text{loc}}(Q)$ and $Lu = 0$ a.e. in Q.

We say that a function $u \in W^{1,2}_{\text{loc}}(Q)$ tends weakly to 0 at a point $x_\circ \in \partial Q$ if $u(x_\delta(x))$ converges to 0 as $\delta \to 0$ weakly in $L^1(\partial Q)$ in some neighborhood $U(x_\circ) \subset \partial Q$ of this point, i.e., for every $h \in L^\infty(U(x_\circ))$

$$\lim_{\delta\to 0} \int_{U(x_\circ)} u(x_\delta(x))h(x)\, dS_x = 0 \tag{6.4.}$$

Lemma 6.2. *Let $p \in (\frac{n}{n-1}, \infty)$ and let $E \subset \partial Q$ be a closed set of finite Hausdorff measure of order $n-q$, $\frac{1}{p} + \frac{1}{q} = 1$. Suppose that $u \in W^{1,2}_{\text{loc}}(Q)$ is a solution of the equation $Lu = 0$, tending weakly to 0 at each point of the set $\partial Q - E$. If $u \in L^1(Q)$, then for each positive number ϵ and any function $v \in C^2(\bar{Q})$ vanishing on ∂Q, the equality*

$$\int_Q L^*(v(x)(1 - \sum_{l=1}^{m} \xi_l(x)))u(x)\,dx = 0$$

holds, where $\{\xi_l\}$ $(l = 1, ...m)$ is the family of functions constructed for ϵ in Lemma 6.1.

PROOF. Let ρ be the function defined in Section 1.4 of Chapter 1. Let $x_{\delta_\circ}$ be an arbitrary point of $\partial Q_{\delta_\circ}$. We introduce a local coordinate system $O_{y_1,...,y_n}$ such that $x_{\delta_\circ}$ is the origin, and the normal to ∂Q_δ at this point has the direction of the axis $0y_n$. The coordinates of the point x in the local system are denoted by $(y_1, ..., y_{n-1}, y_n) = (y', y_n)$ and the function $\rho(x)$ in the local system by $\tilde{\rho}(y)$.
We now consider a function $R(\delta, y', y_n) = \tilde{\rho}(y) - \delta$. For fixed $\delta \in [0, \delta_\circ]$, the 0-level surface of this function coincides with ∂Q_δ. Since

$$\frac{\partial R(0, 0, \delta_\circ)}{\partial y_n} = \frac{\partial \rho(x_\circ)}{\partial \nu} = 1,$$

where $x_\circ = x_{\delta_\circ} + \delta_\circ \nu_{\delta_\circ}(x_{\delta_\circ})$ and $\frac{\partial \rho(x_\delta)}{\partial \nu}$ is the directional derivative in the normal direction ν at $x_\circ \in \partial Q$, it follows from the implicit function theorem that there are positive numbers κ and $\delta_1 \leqq \frac{\delta_\circ}{2}$ such that for $\delta \in [0, \delta_1]$ the connected portion Γ_δ of ∂Q_δ lying on $\partial Q_\delta \cap \{y;\ |y'| < \kappa, y_n > 0\}$ and containing the point with local coordinates $y' = 0, y_n = \delta_\circ - \delta$ is described by the equation $y_n = \varphi(\delta, y')$, where $\varphi \in C^2([0, \delta_1] \times \{|y'| \leqq \kappa\})$. We may assume that $\kappa < \frac{\delta_\circ}{2}$. Thus the surface Γ_δ, $0 \leqq \delta \leqq \delta_1$ does not intersect the hyperplane $y_n = 0$, i.e., $\varphi(\delta, y') > 0$ for $|y'| \leqq \kappa$ and $0 \leqq \delta \leqq \delta_1$. Moreover, we suppose that κ is so small that for some h the cylinder $\Omega^h = \{y;\ |y'| < \kappa, 0 < y_n < \varphi(0, y') + h\}$ for any $\delta \in [0, \delta_1]$ does not contain points of ∂Q_δ different from points of Γ_δ described by the equation $y_n = \varphi(\delta, y')$ for $|y'| < \kappa$.
Since $x_{\delta_\circ}$ was an arbitrary point of ∂Q_δ, by virtue of compactness of $\bar{Q} - Q_{\delta_\circ}$ we can find a finite covering $\Omega_1^{h_1}, ..., \Omega_k^{h_k}$ of the set $\bar{Q} - Q_{\frac{\delta_\circ}{2}}$. It is clear that a family $\{\Omega_1^{h_1}, ..., \Omega_k^{h_k}, Q_{\frac{\delta_\circ}{2}}\}$ is an open covering of $\bar{Q}$. Let $\{\alpha_i(x)\}$ $(i = 0, ..., k)$ be partition of unity subordinate to this covering, i.e., $\alpha_i \in C^\infty(R_n)$ $(i = 0, ..., k)$, $\sum_{i=0}^k \alpha_i(x) = 1$ for $x \in \bar{Q}$, supp $\alpha_\circ \subset Q_{\frac{\delta_\circ}{2}}$ and supp $\alpha_i \subset \Omega_i^{h_i}$ $(i = 1, ..., k)$. We put

$$w(x) = v(x)(1 - \sum_{l=1}^{m} \xi_l(x))$$

and consider the folowing integral

$$\int_Q L^*(w(x))u(x)\,dx = \int_Q L^*(w(x)\sum_{i=0}^{k} \alpha_i(x))u(x)\,dx$$

$$= \sum_{i=0}^{k} \int_Q L^*(w(x)\alpha_i(x))u(x)\,dx.$$

Since supp $\alpha_\circ \subset Q_{\frac{\delta_0}{2}}$, integrating by parts we obtain

$$\int_Q L^*(w(x)\alpha_\circ(x))u(x)\,dx = \int_{Q_{\frac{\delta_0}{2}}} \Big[-D_i(a^{ji}(x)D_j(w\alpha_\circ) - d^i(x)w\alpha_\circ)$$

$$+ b^i(x)D_i(w\alpha_\circ) + c(x)w\alpha_\circ\Big]u\,dx$$

$$= \int_{Q_{\frac{\delta_0}{2}}} \Big[a^{ij}(x)D_iuD_j(\alpha_\circ w) + b^i(x)uD_i(\alpha_\circ w) + d^i(x)D_iu\alpha_\circ w$$

$$+ c(x)w\alpha_\circ u\Big]\,dx = 0.$$

The last integral is equal to 0, since u is a weak solution of $Lu = 0$.
Let Ω^h be one of the cylinders $\{\Omega_1^{h_1}, ..., \Omega_k^{h_k}\}$ and let $\alpha(x)$ be the corresponding element in partition of unity. We put

$$\Omega = \Omega^h \cap \bar{Q} = \{|y'| < \kappa, 0 < y_n \leqq \varphi(0, y')\},$$

where κ is the radius of the base of Ω^h and $\varphi \in C^2([0,\delta_1] \times \{|y'| \leqq \kappa\})$. The surface $\Gamma_\delta = \partial Q_\delta \cap \Omega$ for $\delta \in [0, \delta_1]$ is given by the equation $y_n = \varphi(\delta, y')$, $|y'| < \kappa$. We show that

$$\int_Q L^*(w(x)\alpha(x))u(x)\,dx = 0.$$

Let $0 < \delta \leqq \delta_1$ and put $\Omega_\delta = \{|y| < \kappa;\ 0 < y_n \leqq \varphi(\delta, y')\}$. By A_δ we denote the reflection of the cylinder Ω_δ in Ω defined as follows: to the point $x \in \Omega_\delta$ with local coordinates (y', y_n) we assign the point $A_\delta(x) \in \Omega$ with coordinates

$$\left(y', \frac{y_n}{\varphi(\delta, y')}\varphi(0, y')\right).$$

It is clear that the surface Γ_δ is transformed into $\Gamma_\circ \subset \partial Q$. It is easy to see that

$$\lim_{\delta\to 0} w(A_\delta(x)) = w(x), \quad \lim_{\delta\to 0} D_iw(A_\delta(x)) = D_iw(x)$$

and

$$\lim_{\delta\to 0} D_{ij}w(A_\delta(x)) = D_{ij}w(x)$$

for every $x \in \Omega$ $(i, j = 1, ..., n)$. Let us now observe that

$$\max_{\{x\in Q\}} |\chi_\delta(x)L^*(w(A_\delta(x))\alpha(x))| \leqq b,$$

where b is a positive constant independent of δ and where χ denotes the characteristic function of Ω_δ. Consequently by the Lebesgue Dominated Convergence Theorem we have

$$\lim_{\delta\to 0} \int_\Omega \chi_\delta(x)L^*(w_\delta(x)\alpha(x))u(x)\,dx = \int_\Omega L^*(w(x)\alpha(x))u(x)\,dx,$$

where $w_\delta(x) = w(A_\delta(x))$. We now consider the integral

$$\int_{\Omega_\delta} L^*\big(w_\delta(x)\alpha(x)\big)u(x)\,dx.$$

The function $w_\delta\alpha$, defined on Ω_δ, is equal to 0 on the boundary $\partial\Omega_\delta$ and extending it by 0 outside Ω_δ, we obtain a function belonging to $W^{1,2}(Q)$ with compact support in Q. Therefore using $w_\delta\alpha$ as a test function and integrating by parts we obtain

$$\int_{\Omega_\delta} L^*\big(w_\delta(x)\alpha(x)\big)u(x)\,dx = -\int_{\Gamma_\delta} a^{ij}(x)D_i\big(w_\delta(x)\alpha(x)\big)\nu_j(x)u(x)\,dS_\delta,$$

where $\nu(x) = (\nu_1(x), ..., \nu_n(x))$ denotes the outward unit normal vector to ∂Q_δ. To complete the proof it suffices to show that

$$\lim_{\delta\to 0}\int_{\Gamma_\delta} a^{ij}(x)D_i\big(w_\delta(x)\alpha(x)\big)\nu_j(x)u(x)\,dS_\delta = 0.$$

Let $\widetilde{\Gamma_\delta} = x_\delta^{-1}(\Gamma_\delta)$ (for the definition of the mapping x_δ see Section 1.3 of Chapter 1) and performing a change of variables we obtain

$$\int_{\Gamma_\delta} a^{ij}(x)D_i\big(w_\delta(x)\alpha(x)\big)\nu_j(x)u(x)\,dS_\delta$$
$$= \int_{\widetilde{\Gamma_\delta}} |J_\delta(x)|a^{ij}(x_\delta(x))D_i\big(w_\delta(x_\delta(x))\alpha(x_\delta(x))\big)\nu_j(x_\delta))u(x_\delta(x))\,dS,$$

where $J_\delta(x)$ denotes the Jacobian of the mapping x_δ. Assume that $x \in \widetilde{\Gamma_\delta} \cap \frac{5}{4}K_l$ for certain l. Then for $\delta < \frac{a_{k_n}}{16}$ we have $|x_\delta(x) - x| = \delta < \frac{a_{k_l}}{16}$ (we assume that $a_{k_1} \geqq a_{k_2} \geqq ... \geqq a_{k_m}$) and moreover

$$|A_\delta(x_\delta(x)) - x_\delta(x)| = |\varphi(0, y') - \varphi(\delta, y')|,$$

where (y', y_n) are the local coordinates of the point $x_\delta(x) \in \Gamma_\delta$ satisfying the relation $y_n = \varphi(\delta, y')$, $|y'| < \kappa$. By the continuity of φ we may assume that

$$|\varphi(0, y') - \varphi(\delta, y')| < \frac{a_{k_n}}{16}$$

for all $y' \in \{|y'| \leqq \kappa\}$ provided δ is sufficiently small. Consequently there is a $\delta_2 > 0$ so small that for all $\delta \in (0, \delta_2]$ the point $A_\delta(x_\delta(x))$ belongs to the cube $\frac{3}{2}K_l$, if $x \in \widetilde{\Gamma_\delta} \cap \frac{5}{4}K_l$ $(l = 1, ..., m)$. Thus $|D(w_\delta(x_\delta(x))\alpha(x_\delta(x))| = 0$ for all $\delta \in (0, \delta_2]$, if $x \in \widetilde{\Gamma_\delta} \cap \bigcup_{l=1}^m \frac{5}{4}K_l$, since in this case the point $A_\delta(x_\delta(x))$ belongs to $\bigcup_{l=1}^m \frac{3}{2}K_l$, where $w \equiv 0$. Therefore

$$\lim_{\delta\to 0}\int_{\Gamma_\delta} a^{ij}(x)D_i\big(w_\delta(x)\alpha(x)\big)\nu_j(x)u(x)\,dS_\delta$$
$$= \lim_{\delta\to 0}\int_{\widetilde{\Gamma_\delta} - \bigcup_{l=1}^m \frac{5}{4}K_l} |J_\delta(x)|a^{ij}(x_\delta)D_i\big(w_\delta(x_\delta)\alpha(x_\delta)\big)\nu_j(x_\delta)u(x_\delta)\,dS$$
$$= \lim_{\delta\to 0}\int_{\partial Q - \bigcup_{l=1}^m \frac{5}{4}K_l} \tilde{\chi}_\delta(x)|J_\delta(x)|a^{ij}(x_\delta)D_i\big(w_\delta(x_\delta)\alpha(x_\delta)\big)\nu_j(x)u(x_\delta)\,dS,$$

where $\tilde{\chi}_\delta(x)$ denotes the characteristic function of the set $\widetilde{\Gamma_\delta}$. Here we have used the fact that for $x \in \partial Q$ the normal to ∂Q_δ at $x_\delta(x)$ coincides with the normal to ∂Q at x. At each point of x of the closed set $\partial Q - \bigcup_{l=1}^m \frac{5}{4}K_l$ there exists a neighbourhood $U(x_\circ) \subset \partial Q$ in which $u(x_\delta(x))$ converges weakly to 0 in $L^1(U(x_\circ))$. We select from these neighbourhoods a finite covering $U_1, ..., U_N$. Let $\{\beta_r(x)\}$ $(r = 1, ..., N)$ be partition of unity subordinate to this covering. It is easy to see that

$$(\mathrm{supp}\ \alpha) \cap \widetilde{\Gamma_\delta} = (\mathrm{supp}\ \alpha) \cap \Gamma_\circ = (\mathrm{supp}\ \alpha) \cap \partial Q$$

provided δ is sufficiently small. Therefore

$$\lim_{\delta\to 0} \int_{\partial Q - \bigcup_{l=1}^m \frac{5}{4}K_l} \tilde{\chi}_\delta(x) a^{ij}(x) D_i\big(\alpha(x)w(x)\big)\nu_j(x)u(x_\delta)\,dS$$
$$= \lim_{\delta\to 0} \int_{\partial Q - \bigcup_{l=1}^m \frac{5}{4}K_l} a^{ij}(x) D_i\big(\alpha(x)w(x)\big)\nu_j(x)u(x_\delta)\,dS$$
$$= \sum_{r=1}^N \lim_{\delta\to 0} \int_{U_r} \beta_r(x) a^{ij}(x) D_i\big(w(x)\alpha(x)\big)\nu_j(x)u(x_\delta)\,dS = 0,$$

since

$$\beta_r(x) a^{ij}(x) D_i\big(w(x)\alpha(x)\big)\nu_j(x) \in L^\infty(U_r) \quad (r = 1, ..., N).$$

On the other hand we note that

$$\lim_{\delta\to 0} D_i\big(w_\delta(x_\delta)\alpha(x_\delta)\big) = D_i\big(w(x)\alpha(x)\big) \quad (i = 1, ..., n),$$
$$\lim_{\delta\to 0} a^{ij}(x_\delta) = a^{ij}(x) \quad (i, j = 1, ..., n)$$

and $\lim_{\delta\to 0} J_\delta(x) = 1$ uniformly on ∂Q, Q and ∂Q, respectively. Moreover, it follows from the Banach - Steinhaus theorem that the weak convergence to 0 in L^1 of $u(x_\delta)$ in a neighbourhood of each point of the compact set $\partial Q - \bigcup_{l=1}^m \frac{5}{4}K_l$ implies the boundedness as $\delta \to 0$ of the family of the integrals $\int_{\partial Q - \bigcup \frac{5}{4}K_l} |u(x_\delta(x)|\,dS$. Therefore

$$\lim_{\delta\to 0} \int_{\Gamma_\delta} a^{ij}(x) D_i\big(w_\delta(x)\alpha(x)\big)\nu_j(x)u(x)\,dS_\delta$$
$$= \lim_{\delta\to 0} \int_{\partial Q - \bigcup_{l=1}^m \frac{5}{4}K_l} \tilde{\chi}_\delta(x)u(x_\delta)\Big[|J_\delta(x)|a^{ij}(x_\delta)D_i\big(w_\delta(x_\delta)\alpha(x_\delta)\big)$$
$$- a^{ij}(x)D_i\big(w(x)\alpha(x)\big)\Big]\nu_j(x)\,dS$$
$$+ \lim_{\delta\to 0} \int_{\partial Q - \bigcup_{l=1}^m \frac{5}{4}K_l} \tilde{\chi}_\delta(x) a^{ij}(x) D_i\big(w(x)\alpha(x)\big)\nu_j(x)u(x_\delta)\,dS = 0$$

and this completes the proof.

Lemma 6.3. *Let $p \in (\frac{n}{n-1}, \infty)$ and let $E \subset \partial Q$ be a closed set of finite Hausdorff measure of order $n - q$, $\frac{1}{p} + \frac{1}{q} = 1$. Suppose that u is a solution in $W^{1,2}_{\text{loc}}(Q)$ of the equation $Lu = 0$, tending weakly to 0 at each point of the set $\partial Q - E$. If $u \in L^p(Q)$, then*

$$\int_Q L^*(v(x))u(x)\,dx = 0$$

for every function $v \in C^2(\bar{Q})$ such that $v = 0$ on ∂Q.

PROOF. Let $\epsilon > 0$. We construct for ϵ a covering of E by cubes $\{K_l\}$ $(l = 1, ..., m)$ and corresponding functions $\{\xi_l\}$ satisfying the conditions of Lemma 6.1. By Lemma 6.2 we get

$$\int_Q L^*(v(x))u(x)\,dx = \int_Q L^*(v(x)\sum_{l=1}^m \xi_l(x))u(x)\,dx.$$

We note that supp $\sum_{l=1}^m \xi_l \subset \bigcup_{l=1}^m 2K_l$. Using Hölder's inequality we obtain the estimate

(6.5)

$$\begin{aligned}|\int_Q L^*(v(x))u(x)\,dx| &= |\int_{Q\cap\bigcup_{l=1}^m 2K_l} L^*(v(x)\sum_{l=1}^m \xi_l(x))u(x)\,dx| \\ &\le \Big[\int_{Q\cap\bigcup_{l=1}^m 2K_l} |L^*(v(x)\sum_{l=1}^m \xi_l(x))|^q\,dx\Big]^{\frac{1}{q}} \Big[\int_{Q\cap\bigcup_{l=1}^m 2K_l} |u(x)|^p\,dx\Big]^{\frac{1}{p}}.\end{aligned}$$

Since $\lim_{\epsilon\to 0} |\bigcup_{l=1}^m 2K_l| = 0$, we have

$$\lim_{\epsilon\to 0}\int_{Q\cap\bigcup_{l=1}^m 2K_l} |u(x)|^p\,dx = 0.$$

Therefore to complete the proof it suffices to show that the first integral on the right side of the inequality (6.5) is bounded independently of ϵ. Now consider

$$\begin{aligned}L^*(v\sum_{l=1}^m \xi_l) &= -D_j a^{ij} D_i v \sum_{l=1}^m \xi_l - D_j a^{ij} v D_i(\sum_{l=1}^m \xi_l) \\ &- a_{ij} D_{ij} v(\sum_{l=1}^m \xi_l) - a^{ij} D_i v D_j(\sum_{l=1}^m \xi_l) - a^{ij} D_j v D_i(\sum_{l=1}^m \xi_l) \\ &- a^{ij} v D_{ij}(\sum_{l=1}^m \xi_l) - d^i v D_i(\sum_{l=1}^m \xi_l) - d^i D_i v \sum_{l=1}^m \xi_l - D_i d^i v \sum_{l=1}^m \xi_l \\ &+ b^i D_i v \sum_{l=1}^m \xi_l + b^i v D_i(\sum_{l=1}^m \xi_l) + cv\sum_{l=1}^m \xi_l.\end{aligned}$$

It is obvious that the terms not containing derivatives of $\sum_{l=1}^m \xi_l$ are bounded. We show that the integrals of the form

$$\int_{Q\cap\bigcup_{l=1}^m 2K_l} |D_i(\sum_{r=1}^m \xi_r)|^q\,dx \text{ and } \int_{Q\cap\bigcup_{l=1}^m 2K_l} |v(x)D_{ij}(\sum_{r=1}^m \xi_r)|^q\,dx$$

are bounded. To prove this we decompose $\sum_{l=1}^{m} 2K_l$ into nonintersecting sets:

$$V_1 = 2K_1 - \bigcup_{r=2}^{m} 2K_r,\ V_2 = 2K_2 - \sum_{r=3}^{m} 2K_r, \ldots,$$
$$V_{m-1} = 2K_{m-1} - 2K_m,\ V_m = 2K_m.$$

Note that for each $l = 1, \ldots, m$ we have $\sum_{r=1}^{m} \xi_r(x) = \sum_{r=1}^{l} \xi_r(x)$ for $x \in V_l$. Therefore, using Lemma 6.1, we obtain

$$\int_{Q\cap \bigcup_{l=1}^{m} 2K_l} |D_i(\sum_{r=1}^{m} \xi_r(x))|^q\, dx = \sum_{l=1}^{m} \int_{Q\cap V_l} |D_i(\sum_{r=1}^{l} \xi_r(x))|^q\, dx$$
$$\leqq C^q \sum_{l=1}^{m} \frac{1}{a_{k_l}^q} \int_{V_l} dx \leqq 2^n C^q \sum_{l=1}^{m} a_{k_l}^{n-q},$$

where C is the constant from Lemma 6.1. Since $v = 0$ on ∂Q and $K_l \cap E \neq \varnothing$ for every l, it follows that

$$|v(x)| = \max_{\bar{Q}} |Dv(x)| 2\sqrt{n} a_{k_l}$$

for $x \in 2K_l \cap Q$. Consequently

$$\int_{Q\cap \bigcup_{l=1}^{m} 2K_l} |v(x) D_{ij}(\sum_{r=1}^{m} \xi_r(x))|^q\, dx = \sum_{l=1}^{m} \int_{Q\cap V_l} |v(x) D_{ij}^2(\sum_{r=1}^{l} \xi_r(x))|^q\, dx$$
$$\leqq 2^n \left(C \max_{\bar{Q}} |Dv(x)| 2\sqrt{n}\right)^q \sum_{l=1}^{m} a_{k_l}^{n-q},$$

where C is the constant from Lemma 6.1. It follows from property (iii) of the family $\{K_l\}$ (see Section 6.1) that $\sum_{l=1}^{m} a_{k_l}^{n-q}$ is bounded independently of ϵ and this completes the proof.

Theorem 6.1. *Let $p \in (\frac{n}{n-1}, \infty)$ and let $E \subset \partial Q$ be a closed set of finite Hausdorff measure of order $n - q$, $\frac{1}{p} + \frac{1}{q} = 1$. Suppose that u is a solution in $W^{1,2}(Q) \cap L^p(Q)$ of the equation $Lu = 0$ and that 0 is not eigenvalue of the Dirichlet problem for the operator L. If, in some neighborhood $U(x_\circ)$ of each point $x_\circ \in \partial Q - E$,*

$$\lim_{\delta\to 0} \int_{U(x_\circ)} u(x_\delta(x))^2\, dS_x = 0,$$

then $u \equiv 0$ on Q.

PROOF. It is obvious that the function $u(x)$ converges weakly to 0 at each point of $\partial Q - E$. It follows from Lemma 6.3 that

$$\int_Q L^*(v(x)) u(x)\, dx = 0 \tag{6.6}$$

for all $v \in C^2(\bar{Q})$ with $v(x) = 0$ on ∂Q. The set of such functions is dense in $\overset{\circ}{W}{}^{2,q}(Q)$. Therefore (6.6) holds for all $v \in \overset{\circ}{W}{}^{2,q}(Q)$. Since 0 is not eigenvalue of the Dirichlet problem

$$L^*v = g \text{ in } Q \text{ and } v = 0 \text{ on } \partial Q \tag{6.7}$$

in $\overset{\circ}{W}{}^{2,q}(Q)$, hence for each $g \in L^q(Q)$ there exists a unique solution in $\overset{\circ}{W}{}^{2,q}(Q)$ of the problem (6.7) (see [**KO**])). Consequently

$$\int_Q g(x)u(x)\,dx = 0$$

for every $g \in L^q(Q)$ and $u = 0$ a.e. on Q.

Remark. It is clear from the proof that the convegence of $u(x_\delta)$ to 0 in $L^2(\partial Q)$ may be replaced by the weak convergence in $L^1(U(x_\circ))$ for each $x_\circ \in \partial Q - E$. This means that if a closed set $E \subset \partial Q$ is of finite Hausdorff measure of order $n - q$, then the space $L^p(Q)$ is the uniqueness class for the Dirichlet problem with the boundary condition understood in the weak sense.

6.3. Counter-examples.

Theorem 6.1 remains true in the case $p = \frac{n}{n-1}$, provided the set E consists of finite number of points. If $p < \frac{n}{n-1}$, then the Poisson kernel shows that $L^p(Q)$ is not the uniqueness class for the Dirichlet problem even if the set E consists of a single point.

The following theorem shows that the order of Hausdorff measure in Theorem 6.1 is optimal.

In the proof we need the following result on Hausdorff measure.

Let E be a closed set on the boundary of the unit ball $B(0,1) = \{x;\ |x| < 1\}$ of positive s-dimensional measure with $s \in (n-q, n-1]$. Then there exists a non-negative measure μ with support in E such that

$$\mu(E) = 1 \text{ and } \mu(S) \leqq \text{Const}\, r^s$$

for all balls S of radius r (see [**CA**], Theorem 1, p. 7). In Section 9.1 of Chapter 9 we shall prove a more general version of this result.

Theorem 6.2. *Let $q \in (1,n)$ and $s \in (n-q, n-1]$. Then for any closed set $E \subset \partial B(0,1)$ of positive s-dimensional Hausdorff measure there is a non-zero harmonic function $u(x)$ on $B(0,1)$, belonging to $L^p(B(0,1))$ such that $u \in C(\overline{B(0,1)} - E)$ and $u(x) = 0$ on $\partial B(0,1) - E$.*

PROOF. We define a harmonic function on $B(0,1)$ by

$$u(x) = \int_E \frac{1 - |x|^2}{|x-y|^n}\,d\mu(y).$$

It is clear that $u \in C(\overline{B(0,1)} - E)$, $u(0) = 1$ and $u(x) = 0$ on $\partial B(0,1) - E$. We show that $u \in L^p(B(0,1))$, where $\frac{1}{p} + \frac{1}{q} = 1$. Let g be a positive step function such that

$$\int_{B(0,1)} g(x)^q \, dx = 1.$$

We put $t = \frac{s}{p} + \frac{n-q}{q}$, so that $n - q < t < s$. We introduce the complex variable $z = \zeta + i\tau$, $0 \leqq \zeta \leqq 1$, and consider the function

$$F(z) = \int_{B(0,1)} \int_E \frac{g(x)^{q(1-z)}(1 - |x|^2)}{|x-y|^{t+(n-s)z+1}} \, d\mu(y) \, dx.$$

We show that $|F(i\tau)| \leqq M$ and $|F(1 + i\tau)| \leqq M$, where M is a positive constant independent of $g(x)$. For this we first note that $1 - |x|^2 \leqq 2|x - y|$ for $|x| < 1$ and $|y| = 1$. Therefore

$$\begin{aligned}
|F(i\tau)| &\leqq \int_{B(0,1)} \int_E \frac{g(x)^q(1 - |x|^2)}{|x-y|^{t+1}} \, d\mu(y) \, dx \leqq 2 \int_{B(0,1)} g(x)^q \int_E \frac{d\mu(y)}{|x-y|^t} \, dx \\
&\leqq 2 \int_{B(0,1)} g(x)^q \sum_{k=-1}^{\infty} \int_{2^{-k-1} \leqq |x-y| \leqq 2^{-k}} \frac{d\mu(y)}{|x-y|^t} \, dx \\
&\leqq 2 \int_{B(0,1)} g(x)^q \sum_{k=-1}^{\infty} 2^{t(k+1)} \int_{|x-y| \leqq 2^{-k}} d\mu(y) \\
&\leqq 2 \int_{B(0,1)} g(x)^q \left[\sum_{k=-1}^{\infty} 2^{t(k+1)} \, \mathrm{Const} \, 2^{-ks} \right] dx \\
&= \mathrm{Const} \; 2^{t+1} \sum_{k=-1}^{\infty} 2^{-k(s-t)}.
\end{aligned}$$

Since $s > t$, this series converges. Furthermore

$$\begin{aligned}
|F(1 + i\tau)| &\leqq \int_{B(0,1)} \int_E \frac{1 - |x|^2}{|x-y|^{t+n-s+1}} \, d\mu(y) \, dx \\
&\leqq 2 \int_{B(0,1)} \int_E \frac{d\mu(y) \, dx}{|x-y|^{t+n-s}} = 2 \int_{B(0,1)} \int_{B(0,1)} \frac{dx}{|x-y|^{t+n-s}} \, d\mu(y).
\end{aligned}$$

The last integral is finite, since $s > t$. Consequently the function $F(z)$ is analytic and bounded in the strip $0 < \Re z < 1$, and on the boundary of the strip this function is bounded by a constant M which does not depend on the choice of $g(x)$. Therefore, by virtue of the maximum principle

$$F(p^{-1}) = \int_{B(0,1)} u(x) g(x) \, dx \leqq M.$$

Since this inequality is valid for any positive step function $g(x)$ such that $\|u\|_q = 1$, it follows that

$$\int_{B(0,1)} u(x)^p\, dx \leqq M^p$$

and this completes the proof.

The following result shows that without additional conditions on the exceptional set on the boundary, the space L^p, $p < \infty$, is not a uniqueness class for the Dirichlet problem in question.

Theorem 6.3. *Let $B(0,1)$ be the unit disc in the plane. There exists a non zero function $u(x)$, harmonic in Q, which belongs to $L^p(B(0,1))$ for all $p < \infty$, is continuous on $\overline{B(0,1)} - E$, where E is closed subset of ∂Q of zero linear Lebesgue measure.*

PROOF. We divide the boundary $\partial B(0,1)$ into eight equal parts, choose four arcs which are not adjacent and denote them by $\gamma_{1,i}$, $i = 1,2,3,4$. We take the arcs to be closed. The construction now proceeds by induction. At the k-th step ($k = 2,3,\ldots$) each of the arcs selected at the preceding step ($i = 1,..,m_{k-1}$), where m_{k-1} is the number of arcs selected at the $(k-1)$-st step is divided into $2^{2^{k-1}+1}$ equal parts. We choose every other one of the arcs obtained by this partition. We denote them by $\gamma_{k,i}$ ($i = 1,\ldots,m_k$) tracing out the boundary in a counterclockwise direction. It is easy to see that $m_k = 2^{2^k}$ and $\gamma_k = \frac{2\pi}{2^{2^k+k}}$, where γ_k is the length of each arc $\gamma_{k,i}$ selected at the k-th step. The arcs $\gamma_{k,i}$ are taken to be closed. Put $E_k = \bigcup_{i=1}^{m_k} \gamma_{k,i}$. Note that E_k is a closed subset of $\partial B(0,1)$. We now consider the sequence of functions

$$u_k(r,\psi) = c_k \int_{E_k} P(r,\psi-\theta)\, d\theta \equiv c_k \int_{E_k} \frac{(1-r^2)\, d\theta}{1+r^2-2r\cos(\psi-\theta)},$$

harmonic in $B(0,1)$, where (r,ψ) are polar coordinates on the plane. The constants c_k are chosen from the condition

$$u_k(0,0) = 2\pi, \quad \text{that is } 2\pi = c_k|E_k| = c_k\gamma_k m_k,$$

hence $c_k = 2^k$. We show that for each $k = 1,2,\ldots$ and any point $(r,\psi) \in B(0,1)$ $(0 \leqq r < 1, 0 \leqq \psi < 2\pi)$ the estimate

$$u_{k+1}(r,\psi) \leqq u_k(r,\psi) + c_k F_{k+1}(r) \tag{6.8}$$

holds, where

$$F_{k+1}(r) = \int_{-\frac{\gamma_{k+1}}{2}}^{\frac{\gamma_{k+1}}{2}} \frac{1-r^2}{1+r^2-2r\cos\theta}\, d\theta.$$

We denote by $\gamma'_{k,i}$ the arc of length γ_k following after $\gamma_{k,i}$ in a counterclockwise traversal of the boundary $\partial B(0,1)$ ($i = 1,\ldots,m_k$). We introduce the notation

$$J_{k,i}(r,\psi) = \int_{\gamma_{k,i}} P(r,\psi-\theta)\, d\theta \text{ and } J'_{k,i}(r,\psi) = \int_{\gamma'_{k,i}} P(r,\psi-\theta)\, d\theta.$$

Let $(1,\psi) \in \gamma_{k+1,1}$ for certain k. It follows the monotonicity of the Poisson kernel $P(r,\psi-\theta)$ that

$$J_{k+1,1}(r,\psi) \leqq F_{k+1}(r),$$
$$J_{k+1,i}(r,\psi) \leqq J'_{k+1,i-1}(r,\psi) \text{ for } 2 \leqq i \leqq \frac{m_{k+1}}{2}+1,$$
$$J_{k+1,i}(r,\psi) \leqq J'_{k+1,i}(r,\psi) \text{ for } \frac{m_{k+1}}{2} \leqq i \leqq m_{k+1}.$$

Adding these inequalities, we obtain

$$\sum_{i=1}^{m_{k+1}} J_{k+1,i}(r,\psi) \leqq \sum_{i=1}^{m_{k+1}} J'_{k+1,i}(r,\psi) + F_{k+1}(r)$$

(here we put on the right side the term $J'_{k+1,\frac{m_{k+1}}{2}+1}(r,\psi)$) . Similarly, the same inequality is obtained in the case $(1,\psi) \in E_{k+1}$. Thus

$$\sum_{i=1}^{m_k} J_{k,i}(r,\psi) = \sum_{i=1}^{m_{k+1}} J_{k+1,i}(r,\psi) + \sum_{i=1}^{m_{k+1}} J'_{k+1,i}(r,\psi)$$
$$\geqq 2\sum_{i=1}^{m_{k+1}} J_{k+1,i}(r,\psi) - F_{k+1}(r),$$

or

$$\sum_{i=1}^{m_{k+1}} J_{k+1,i}(r,\psi) \leqq \frac{1}{2}\sum_{i=1}^{m_k} J_{k,i}(r,\psi) + \frac{1}{2}F_{k+1}(r).$$

Multiplying this latter inequality by $c_{k+1} = 2^{k+1}$, we obtain (6.8). Since (6.8) is valid for all $k = 1, 2, \ldots$. it follows that

$$u_{k+1}(r,\psi) \leqq u_1(r,\psi) + \sum_{i=1}^{k} c_i F_{i+1}(r).$$

We note that $u_1(r,\psi) < 4\pi$, $F_{i+1}(r) < 2\pi$, and also

$$F_{i+1}(r) = \int_{-\frac{\gamma_{i+1}}{2}}^{\frac{\gamma_{i+1}}{2}} \frac{(1-r^2)\,d\theta}{(1-r)^2 + 2r(1-\cos\theta)} < \frac{2\gamma_{i+1}}{1-r}.$$

Therefore

$$u_{k+1}(r,\psi) \leqq 4\pi + \sum_{i=1}^{i_0-1} c_i F_{i+1}(r) + \sum_{i=i_0}^{k} c_i F_{i+1}(r)$$
$$\leqq 4\pi + 2\pi\sum_{i=1}^{i_0-1} 2^i + \frac{1}{1-r}\sum_{i=i_0}^{k} 2^{i+1}\gamma_{i+1}$$
$$\leqq 2\pi 2^{i_0} + \frac{1}{1-r}\sum_{i=i_0}^{\infty} 2^i\gamma_i.$$

It is easy to verify that

$$\sum_{i=i_o+1}^{\infty} 2^i \gamma_i < 2^{i_o} \gamma_{i_o}$$

for any $i_o \geqq 1$. We choose

$$i_o = 1 + \left[\log_2 \log_2 \left(\frac{2\pi}{1-r} \right) \right],$$

where the square brackets denote the integral part. It is easy to see that $\gamma_{i_o} \leqq 1 - r$. Hence we obtain the estimate

$$\begin{aligned} u_{k+1}(r,\psi) &\leqq (2\pi+1)2^{i_o} \leqq (2\pi+1)2^{\log_2 \log_2 \frac{2\pi}{1-r}+1} \\ &\leqq 2(2\pi+1)\log_2 \frac{2\pi}{1-r}, \end{aligned}$$

which is uniform with respect to ψ. By virtue of this estimate, the sequence of functions $u_k(x)$ harmonic in $B(0,1)$ is uniformly bounded in every circle $B_m = B(0, 1 - \frac{1}{m})$, $m = 2, 3, \ldots$, so from this sequence we may select a subsequence which converges uniformly on any subdomain $B'_m \subset B_m$ (with $\bar{B}'_m \subset B_m$). This subsequence is uniformly bounded on B_{m+1}, and so one may select a subsequence which converges uniformly on any subdomain $B'_{m+1} \subset B_{m+1}$ (with $\bar{B}'_{m+1} \subset B_{m+1}$) and so on. From these subsequences we choose a diagonal one which will converge uniformly on any subdomain $B' \subset B(0,1)$ (with $\bar{B}' \subset B(0,1)$) to a function $u(x)$ harmonic in $B(0,1)$. It is clear that

$$0 \leqq u(r,\psi) \leqq 2(2\pi+1)\log_2 \frac{2\pi}{1-r},$$

i.e., $u \in L^p(B(0,1))$ for $p < \infty$. It is also obvious that $u \neq 0$, since $u_k(0) = 2\pi$. Put $E = \bigcap_{k=1}^{\infty} E_k$. The set E is closed subset of $\partial B(0,1)$ and $|E| = 0$ since $|E_k| = \frac{2\pi}{2^k} \to 0$ as $k \to \infty$. For any point $\omega \in \partial B(0,1) - E$, clearly there are numbers $\epsilon > 0$ and k_o such that

$$\{\psi \in \partial B(0,1);\ |\psi - \omega| < \epsilon\} \cap E_{k_o} = \emptyset.$$

Since $E_{k+1} \subset E_k$ for all $k > k_o$, we see that

$$\{\psi \in \partial B(0,1);\ |\psi - \omega| < \epsilon\} \cap E_k = \emptyset.$$

We take a neighbourhood $U(\omega) = \{\psi \in \partial B(0,1);\ |\psi - \omega| < \frac{\epsilon}{2}\}$ of the point ω on $\partial B(0,1)$. Then for all $\psi \in U(\omega)$, all $r \in (0,1)$ and all $k \geqq k_o$ the estimate

$$\begin{aligned} u_k(r,\psi) =& c_k \int_{E_k} \frac{1-r^2}{(1-r)^2 + 4r\sin^2\left(\frac{\psi-\theta}{2}\right)}\, d\theta \\ &\leqq c_k |E_k| \frac{1-r^2}{4r\sin^2\frac{\epsilon}{4}} \leqq \frac{\pi}{\sin^2\frac{\epsilon}{4}} \frac{1-r}{r} \end{aligned}$$

holds. This estimate is also satisfied by the limit function $u(r,\psi)$. Therefore the function $u(r,\psi)$ converges to zero as $r \to 1$, uniformly with respect to $\psi \in U(\omega)$. This completes the proof.

CHAPTER 7

APPLICATIONS OF THE L^2-METHOD

In this chapter we show how the L^2-approach to the Dirichlet problem for uniformly elliptic equations, developed in Chapter 3, can be used to solve the Dirichlet problem for degenerate elliptic equations. We are concerned here with elliptic equations with degeneracy on the boundary. It is clear from results of Chapter 3, that uniform ellipticity is an essential assumption which allows us to recover L^2-boundary data. We show that the lack of uniform ellipticity can be compensated by suitable assumptions on the coefficients d^i. Finally, we briefly discuss the application of our method to a non-local problem in the sense of Bitsadze–Samarskii.

7.1 Energy estimate for degenerate elliptic equations.

In Sections 7.1–7.4 of this chapter we are concerned with the Dirichlet problem for a degenerate elliptic equation

$$Eu + \lambda u = -D_i\big(\rho(x)a^{ij}(x)D_j u\big) + d^i(x)D_i u + c(x)u + \lambda u = f(x) \text{ in } Q \tag{7.1}$$

$$u(x) = \varphi(x) \text{ on } \partial Q, \tag{7.2}$$

where λ is a real parameter and $\rho(x)$ is C^2–function on $\bar{Q}$ equivalent to the distance for $x \in \bar{Q}$ and its properties are described in Sections 1.3–1.4.

Throughout Sections 7.1–7.4 we make the following assumptions on E

(H_1) The coefficients of E are in $C^\infty(\bar{Q})$, $a^{ij} = a^{ji}$ $(i, j = 1, ..., n)$ and $c(x) \geqq \beta$ on Q for some constant $\beta > 0$.

(H_2) There exists a constant $\gamma >$ such that

$$\gamma^{-1}|\xi|^2 \leqq a^{ij}(x)\xi_i\xi_j \leqq \gamma|\xi|^2$$

for all $x \in Q$ and $\xi \in R_n$.

(H_3) $f \in L^2(Q)$.

Since the elliptic equation (7.1) degenerates on ∂Q, the theory of second–order equations with non–negative characteristic form asserts that the boundary condition is to imposed on a certain subset of ∂Q, which can be described with the aid of the so called Fichera function ([**OR**] p.17). In our situation the Fichera function is reduced to $z(x) = d^i(x)D_i\rho(x)$. Consequently, following the terminology of [**OR**], the boundary condition (7.2) should be imposed on

$$\Sigma_2 = \{x \in \partial Q;\ d^i(x)D_i\rho(x) > 0\}.$$

Here we assume that

(H_4) $d^i(x)D_i\rho(x) > 0$ on ∂Q,

therefore $\Sigma_2 = \partial Q$.

We need the following result due to Langlais ([**LA, Theorem 2.7**]).

Theorem 7.1. *Let $f \in W^{l,\infty}(Q)$ with $l \geqq 1$. Then there exists κ, $0 < \kappa < 1$, with $\kappa < \inf_{\partial Q} d^i(x) D_i\rho(x)$, such that any solution $u \in C^2(Q) \cap C(\bar{Q})$ of (7.1),(7.2) (with $\varphi(x) = 0$ on ∂Q) satisfies the estimate*

$$\sup_Q |\rho(x)^{1-\kappa} Du(x)| \leqq C(l) \|f\|_{l,\infty},$$

where $C(l) > 0$ is a constant.

Also, according to [**LA, Theorem 2.3**] the problem (7.1),(7.2) is solvable in $C^2(Q) \cap C(\bar{Q})$ for each $\varphi \in C(\partial Q)$ and $\lambda > 0$.

To construct a solution of (7.1),(7.2) in $W^{1,2}_{\text{loc}}(Q)$ we need the following form of the energy estimate.

Lemma 7.1. *Let $\{\varphi_m\}$ and $\{f_m\}$ be sequences in $C^2(\partial Q)$ and $C^1(\bar{Q})$, respectively, such that*

$$\lim_{m\to\infty} \int_{\partial Q} [\varphi_m(x) - \varphi(x)]^2 dS_x = 0 \text{ and } \lim_{m\to\infty} \int_Q [f_m(x) - f(x)]^2 dx = 0.$$

Suppose that u_m is a solution of (7.1), with $f = f_m$, in $C^2(Q) \cap C(\bar{Q})$, satisfying the boundary condition

$$u_m(x) = \varphi_m(x) \text{ on } \partial Q. \tag{7.2m}$$

Then there exist positive constants $\lambda_\circ$ and C, independent of m, such that

$$\begin{aligned} &\int_Q |D^2 u_m(x)|^2 \rho(x)^3\, dx + \int_Q |Du_m(x)|^2 \rho(x)\, dx + \int_Q u_m(x)^2\, dx \\ &\quad \leqq C\Big(\int_{\partial Q} f_m(x)^2 + \int_{\partial Q} \varphi(x)^2\, dS_x\Big), \end{aligned} \tag{7.4}$$

for all $m = 1, 2, \ldots$ and $\lambda \geqq \lambda_\circ$.

PROOF. According to Theorem 7.1 and the remark following this theorem, for each m there exists a solution u_m of (7.1),(7.2m) in $C^2(Q) \cap C(\bar{Q})$ with $\rho^{1-\kappa} Du_m \in L^\infty(Q)$ provided $\lambda > 0$. Multiplying (7.1) by u_m and integrating by parts we obtain

$$\begin{aligned} &\delta \int_{\partial Q_\delta} a^{ij} D_i u_m u_m D_j \rho\, dS_x + \int_{Q_\delta} \rho a^{ij} D_i u_m D_j u_m\, dx \\ &\quad + \int_{Q_\delta} d^i D_i u_m u_m dx + \int_{Q_\delta} c u_m^2\, dx + \lambda \int_{Q_\delta} u_m^2\, dx = \int_{Q_\delta} f_m u_m\, dx. \end{aligned} \tag{7.5}$$

The first integral can be estimated using Young's inequality

$$\Big|\int_{\partial Q_\delta} \delta a^{ij} D_i u_m u_m D_j \rho dS_x\Big| \leqq C_1 \delta^2 \int_{\partial Q_\delta} |Du_m|^2\, dS_x + \int_{\partial Q_\delta} u_m^2\, dS_x, \tag{7.6}$$

where C_1 is independent of δ. Integrating the third integral by parts we get

$$\int_{Q_\delta} d^i D_i u_m u_m \, dx = \frac{1}{2}\int_{Q_\delta} d^i D_i(u_m^2)\, dx = -\frac{1}{2}\int_{\partial Q_\delta} d^i D_i \rho u_m^2 \, dS_x - \frac{1}{2}\int_{Q_\delta} D_i d^i u_m^2 \, dx. \tag{7.7}$$

Combining (7.5),(7.6),(7.7) with (H_2) we arrive at the estimate

$$\begin{aligned}\gamma^{-1}&\int_{Q_\delta} \rho |Du_m|^2 \, dS_x + \int_{Q_\delta} (\lambda - \frac{1}{2} + c - \frac{1}{2} D_i d^i) u_m^2 \, dx \\ &\leqq C_1 \delta^2 \int_{\partial Q_\delta} |Du_m|^2 \, dS_x + \int_{\partial Q_\delta} (\frac{1}{2} d^i D_i \rho + 1) u_m^2 \, dS_x + \frac{1}{2}\int_{Q_\delta} f_m^2 \, dx.\end{aligned}$$

Since $\rho^{1-\kappa} Du_m \in L^\infty(Q)$, $\lim_{\delta\to 0} \delta^2 \int_{\partial Q_\delta} |Du_m|^2 \, dS_x = 0$. Consequently taking λ sufficiently large, say $\lambda \geqq \lambda_\circ$, and letting $\delta \to 0$, we get

$$\int_Q \rho |Du_m|^2 \, dx + \int_Q u_m^2 \, dx \leqq C_2 \Big(\int_{\partial Q} \varphi_m^2 \, dS_x + \int_Q f_m^2 \, dx \Big) \tag{7.8}$$

for all m, where $C_2 > 0$ is a constant independent of m. To estimate the integral $\int_Q |D^2 u_m|^2 \rho^3 \, dx$, we first observe that, if $v \in W^{2,2}(Q)$ and has compact support in Q, then

$$\int_Q \rho a^{ij} D_i u_m D_{jk} v \, dx + \int_Q d^i D_i u_m D_k v \, dx + \int_Q (c+\lambda) u_m D_k v \, dx = \int_Q f_m D_k v \, dx.$$

Integrating the first integral by parts we get

$$\begin{aligned}&\int_Q D_k \rho a^{ij} D_i u_m D_j v \, dx + \int_Q \rho D_k a^{ij} D_i u_m D_j v \, dx + \int_Q \rho a^{ij} D_{ki} u_m D_j v \, dx \\ &- \int_Q d^i D_i u_m D_k v \, dx - \int_Q (c+\lambda) u_m D_k v \, dx = - \int_Q f_m D_k v \, dx.\end{aligned}$$

Letting $v = D_k u_m (\rho - \delta)^2$ on Q_δ and $v = 0$ on $Q - Q_\delta$ we deduce from the last equation

(7.9)

$$\begin{aligned}&\int_{Q_\delta} D_k \rho a^{ij} D_i u_m D_{jk} u_m (\rho-\delta)^2 \, dx + 2\int_{Q_\delta} D_k \rho a^{ij} D_i u_m D_k u_m D_j \rho (\rho - \delta)\, dx \\ &+ \int_{Q_\delta} \rho D_k a^{ij} D_i u_m D_{jk} u_m (\rho-\delta)^2 \, dx + 2\int_{Q_\delta} \rho D_k a^{ij} D_i u_m D_k u_m (\rho-\delta) D_j \rho \, dx \\ &+ \int_{Q_\delta} \rho a^{ij} D_{ki} u_m D_{kj} u_m (\rho-\delta)^2 \, dx + 2\int_{Q_\delta} \rho a^{ij} D_{ki} u_m D_k u_m (\rho-\delta) D_j \rho \, dx \\ &- \int_{Q_\delta} d^i D_i u_m D_{kk} u_m (\rho-\delta)^2 \, dx - 2\int_{Q_\delta} d^i D_i u_m D_k u_m (\rho-\delta) D_k \rho \, dx \\ &- \int_{Q_\delta} (c+\lambda) u_m D_{kk} u_m (\rho-\delta)^2 \, dx - 2\int_{Q_\delta} (c+\lambda) u_m D_k u_m (\rho-\delta) D_k \rho \, dx \\ &= - \int_{Q_\delta} f D_{kk} u_m (\rho-\delta)^2 \, dx - 2\int_{Q_\delta} f D_k u_m (\rho-\delta) D_k \rho \, dx.\end{aligned}$$

Let us denote the integrals on the left side of (7.9) by $J_1, ..., J_{10}$. Estimation of these integrals can be obtained as follows

$$J_5 \geqq \gamma^{-1} \int_{Q_\delta} \sum_{j=1}^{n} |D_{jk} u_m|^2 \rho(\rho-\delta)^2 \, dx. \tag{7.10}$$

Using the Young inequality we get

$$\begin{aligned} |J_1 + J_2 + J_3 + J_4| \leqq C_3(\epsilon) \int_{Q_\delta} |Du_m|^2 (\rho - \delta) \, dx \\ + \epsilon \int_{Q_\delta} \sum_{j=1}^{n} |D_{jk} u_m|^2 (\rho-\delta)^3 \, dx. \end{aligned} \tag{7.11}$$

Similarily we have

$$\begin{aligned} |J_6 + J_7| &\leqq C_4(\epsilon) \Big[\int_{Q_\delta} \rho |Du_m|^2 \, dx + \int_{Q_\delta} |Du_m|^2 (\rho-\delta) \, dx \Big] \\ &+ \epsilon \Big[\int_{Q_\delta} \sum_{j=1}^{n} |D_{kj} u_m|^2 \rho(\rho-\delta)^2 \, dx + \int_{Q_\delta} \sum_{j=1}^{n} |D_{jk} u_m|^2 (\rho-\delta)^3 \, dx \Big], \end{aligned} \tag{7.12}$$

and

$$\begin{aligned} |J_9| + | \int_{Q_\delta} f D_{kk} u_m (\rho-\delta)^2 \, dx| &\leqq C_5(\epsilon) \Big(\int_{Q_\delta} u_m^2 \, dx + \int_{Q_\delta} f^2 \, dx \Big) \\ &+ \epsilon \int_{Q_\delta} \sum_{j=1}^{n} |D_{kj} u_m|^2 (\rho-\delta)^3 \, dx, \end{aligned} \tag{7.13}$$

and finally

$$|J_8 + J_{10}| \leqq C_6 \Big[\int_{Q_\delta} |Du_m|^2 (\rho-\delta) \, dx + \int_{Q_\delta} u_m^2 \, dx \Big], \tag{7.14}$$

where C_i are independent of m and δ. A constant $\epsilon > 0$ is yet to be determined. We deduce from (7.9)–(7.14) that

$$\begin{aligned} &\int_{Q_\delta} ((\gamma^{-1} - \epsilon)\rho(\rho-\delta)^2 - 3\epsilon(\rho-\delta)^3) \sum_{j=1}^{n} |D_{jk} u_m|^2 \, dx \\ &\leqq C_7 \Big[\int_{Q_\delta} |Du_m|^2 (\rho-\delta) \, dx + \int_{Q_\delta} |Du_m|^2 \rho \, dx + \int_{Q_\delta} f^2 \, dx + \int_{Q_\delta} u_m^2 \, dx \Big], \end{aligned}$$

where $C_7 > 0$. Since

$$(\gamma^{-1}-\epsilon)\rho(\rho-\delta)^2 - 3\epsilon(\rho-\delta)^3 = (\rho-\delta)^2\Big[(\gamma^{-1}-\epsilon)\rho - 3\epsilon(\rho-\delta)\Big]$$
$$= (\rho-\delta)^2\Big[(\gamma^{-1}-\epsilon)(\rho-\delta) + \delta(\gamma^{-1}-\epsilon) - 3\epsilon(\rho-\delta)\Big]$$
$$= (\rho-\delta)^2\Big[(\gamma^{-1}-4\epsilon)(\rho-\delta) + \delta(\gamma^{-1}-\epsilon)\Big] > (\rho-\delta)^3(\gamma^{-1}-4\epsilon)$$

for ϵ sufficiently small, say $\epsilon = \frac{\gamma^{-1}}{5}$, the last two inequalities yield

$$\int_{Q_\delta}\sum_{j=1}^{n}|D_{jk}u_m|^2(\rho-\delta)^3\,dx \leqq 5\gamma C_7\Big[\int_{Q_\delta}|Du_m|^2(\rho-\delta)\,dx + \int_{Q_\delta}|Du_m|^2\rho\,dx + \int_{Q_\delta} f^2\,dx + \int_{Q_\delta} u_m^2\,dx\Big]. \tag{7.15}$$

Letting $\delta \to 0$ in (7.15) and combining the resulting inequality with (7.8), we easily arrive at (7.6).

Lemma 7.1 shows that a possible solution to the problem (7.1),(7.2) lies in the space $\widetilde{W}^{2,2}(Q)$ defined by

$$\widetilde{W}^{2,2}(Q) = \{u;\ u \in W^{1,2}_{loc}(Q) \text{ and } \int_Q |D^2u(x)|^2\rho(x)^3\,dx + \int_Q |Du(x)|^2\rho(x)\,dx + \int_Q u(x)^2\,dx < \infty\}$$

and equipped with the norm

$$\|u\|_{\widetilde{W}^{2,2}} = \int_Q |D^2u(x)|^2\rho(x)^3\,dx + \int_Q |Du(x)|^2\rho(x)\,dx + \int_Q u(x)^2\,dx.$$

The proof that u_m converges to solution of (7.1),(7.2) will be given in Section 7.4.

7.2. Traces in $\widetilde{W}^{2,2}(Q)$.

We require first the following property of the space $\widetilde{W}^{2,2}(Q)$.

Lemma 7.2. *If $u \in \widetilde{W}^{2,2}(Q)$, then $\delta^2\int_{\partial Q_\delta}|Du(x)|^2\,dS_x$ is continuous on $[0,\delta_\circ]$ and moreover*

$$\lim_{\delta\to 0}\delta^2\int_{\partial Q_\delta}|Du(x)|^2\,dS_x = 0.$$

PROOF. Let $0 < \delta < \delta_\circ$, then

$$\int_{Q_\delta - Q_{\delta_\circ}} \rho[D_i u(x)]^2\, dx = \int_\delta^{\delta_\circ} \mu\, d\mu \int_{\partial Q_\mu} [D_i u(x)]^2\, dS_x$$

$$= \int_\delta^{\delta_\circ} \mu\, d\mu \int_{\partial Q} [D_i u(x_\mu(x_\circ))]^2 \frac{dS_\mu}{dS_\circ}\, dS_\circ = \frac{\delta_\circ^2}{2} \int_{\partial Q} [D_i u(x_{\delta_\circ}(x_\circ))]^2 \frac{dS_{\delta_\circ}}{dS_\circ}\, dS_\circ$$

$$- \frac{\delta^2}{2} \int_{\partial Q} [D_i u(x_\delta(x_\circ))]^2 \frac{dS_\delta}{dS_\circ}\, dS_\circ$$

$$- \int_\delta^{\delta_\circ} \mu^2 \int_{\partial Q} \Big[\sum_{j=1}^n D_{ji} u(x_\mu(x_\circ)) D_i u(x_\mu(x_\circ)) \frac{\partial x_\mu}{\partial \mu} \frac{dS_\mu}{dS_\circ}$$

$$+ [D_i u(x_\mu(x_\circ))]^2 \frac{\partial}{\partial \mu} \Big(\frac{dS_\mu}{dS_\circ}\Big) \Big]\, dS_\circ.$$

Fom this identity we can compute

$$\delta^2 \int_{\partial Q} [D_i u(x_\delta(x_\circ))]^2 \frac{dS_\delta}{dS_\circ}\, dS_\circ$$

and express this integral in terms of other integrals which are continuous on $[0, \delta_\circ]$, since $u \in \widetilde{W}^{2,2}(Q)$. On the other hand $\lim_{\delta \to 0} \frac{dS_\delta}{dS_\circ} = 1$ uniformly on ∂Q, therefore the continuity of the integral $\delta^2 \int_{\partial Q_\delta} |Du(x)|^2\, dS_x$ easily follows. Assuming that $\lim_{\delta \to 0} \delta^2 \int_{\partial Q_\delta} |Du(x)|^2\, dS_x > 0$, we would have

$$\delta^2 \int_{\partial Q_\delta} |Du(x)|^2\, dS_x > a \text{ on } (0, \delta_1]$$

for some constants $a > 0$ and $\delta_1 > 0$ and this would imply that

$$\int_{Q - Q_{\delta_1}} \rho(x) |Du(x)|^2\, dx = \int_0^{\delta_1} \mu\, d\mu \int_{\partial Q_\mu} |Du(x)|^2\, dS_x = \infty$$

and we get a contradiction.

Lemma 7.3. *Let $u \in \widetilde{W}^{2,2}(Q)$ be a solution of (7.1), then $\int_{\partial Q_\delta} u(x)^2\, dS_x$ is bounded on $(0, \delta_\circ]$.*

PROOF. Multiplying (7.1) by u and integrating over Q_δ we obtain

$$\frac{1}{2} \int_{\partial Q_\delta} u^2 d^i D_i \rho\, dS_x = -\frac{1}{2} \int_{Q_\delta} D_i d^i u^2\, dx + \int_{Q_\delta} \rho a^{ij} D_i u D_j u\, dx$$

$$+ \delta \int_{\partial Q_\delta} a^{ij} D_i u u D_j \rho\, dS_x + \int_{Q_\delta} (c + \lambda) u^2\, dx - \int_{Q_\delta} f u\, dx.$$

We may assume that

$$b = \inf_{Q - Q_{\delta_\circ}} d^i(x) D_i \rho(x) > 0,$$

taking $\delta_\circ$ sufficiently small, if necessary. Let us now observe that by Young's inequality we have

$$\delta \int_{\partial Q_\delta} a^{ij} D_i u u D_j \rho \, dS_x \leqq C\delta^2 \int_{\partial Q_\delta} |Du|^2 \, dS_x + \frac{b}{4} \int_{\partial Q_\delta} u^2 \, dS_x,$$

where $C > 0$ is a constant depending on n, b and $\|a^{ij}\|_\infty$, consequently the result follows from Lemma 7.2.

We are now in a position to state the main result of this section.

Theorem 7.2. *Let $u \in \widetilde{W}^{2,2}(Q)$ be a solution of (7.1). Then there exists a function $\varphi \in L^2(\partial Q)$ such that*

$$\lim_{\delta \to 0} \int_{\partial Q} [u(x_\delta(x)) - \varphi(x)]^2 \, dS_x = 0.$$

PROOF. Since by Lemma 7.3, $\int_{\partial Q} u(x_\delta(x))^2 \, dS_x$ is bounded, there exists a sequence $\delta_m \to 0$, as $m \to \infty$, and a function $\varphi \in L^2(\partial Q)$ such that

$$\lim_{m \to \infty} \int_{\partial Q} u(x_{\delta_m}) g(x) \, dS_x = \int_{\partial Q} \varphi(x) g(x) \, dS_x$$

for each $g \in L^2(\partial Q)$. We prove that above relation remains valid if the sequence $\{\delta_m\}$ is replaced by the parameter δ. Since $\int_{\partial Q} u(x_\delta(x)) g(x) dS_x$ is continuous on $(0, \delta_\circ]$, it suffices to prove the existence of the limit at 0 with g replaced by $\Psi \in C^1(\bar{Q})$. Integrating by parts we obtain

$$\int_{\partial Q_\delta} d^i D_i \rho \Psi u \, dS_x = -\int_{Q_\delta} D_i(d^i \Psi) u \, dx + \int_{Q_\delta} (c + \lambda) \Psi u \, dx$$
$$+ \int_{Q_\delta} \rho a^{ij} D_i u D_j \Psi \, dx + \delta \int_{\partial Q_\delta} a^{ij} D_i u D_j \rho \Psi \, dS_x - \int_{Q_\delta} f \Psi \, dx.$$

Using Lemma 7.2, the continuity of the left side of the above equation easily follows. Letting $\delta \to 0$, we deduce from the last equation that

$$\int_{\partial Q} \varphi \Psi d^i D_i \rho \, dS_x = -\int_Q D_i(d^i \Psi) u \, dx + \int_Q (c + \lambda) \Psi u \, dx$$
$$+ \int_Q \rho a^{ij} D_i u D_j \Psi \, dx - \int_Q f \Psi \, dx \equiv \int_Q F(\Psi) \, dx.$$

It is clear that this relation continues to hold for $\Psi \in W^{1,2}(Q)$. Now taking $\Psi(x) = u(x^\delta(x))$, where $x^\delta : Q \to \bar{Q}_{\frac{\delta}{2}}$ is the mapping defined in Section 3.4, we get

$$\int_{\partial Q} \varphi(x) d^i(x) D_i \rho(x) \, dS_x = \int_{Q_\delta} F(u(x)) \, dx + \int_{Q - Q_\delta} u(x^\delta(x)) \, dx.$$

As in the proof of Theorem 3.4, one can show that

$$\lim_{\delta\to 0}\int_{Q_\delta} F(u(x))\,dx = \lim_{\delta\to 0}\int_{\partial Q} u(x_\delta(x))^2 d^i(x) D_i\rho(x)\,dS_x$$

and

$$\lim_{\delta\to 0}\int_{Q-Q_\delta} F(u(x^\delta(x))\,dx = 0.$$

The result now follows from the uniform convexity of $L^2(\partial Q)$.

7.3. Existence of solutions of the problem (7.1),(7.2).

Theorem 7.2 suggests the following approach to the Dirichlet problem (7.1), (7.2).

Let $\varphi \in L^2(\partial Q)$. A solution u of (7.1) in $\widetilde{W}^{2,2}(Q)$ is a solution of the Dirichlet problem with the boundary condition (7.2) if

$$\lim_{\delta\to 0}\int_{\partial Q}[u(x_\delta(x)) - \varphi(x)]^2\,dS_x = 0. \tag{7.9}$$

Theorem 7.3. *Let $\lambda \geqq \lambda_\circ$ (where $\lambda_\circ$ is a constant from Lemma 7.1). Then for every $\varphi \in L^2(\partial Q)$ there exists a unique solution $u \in \widetilde{W}^{2,2}(Q)$ of the problem (7.1),(7.2).*

PROOF. Let u_m be a sequence of solutions of the problem (7.1),(7.2m) from Lemma 7.1. By the estimate (7.4) there exists a subsequence, which we relabel again as u_m, converging weakly to a function $u \in \widetilde{W}^{2,2}(Q)$. According to Theorem 1.3 we may assume that u_m tends to u in $L^2(Q)$ and a.e. on Q. Evidently u satisfies (7.1). By virtue of Theorem 7.2 there exists a trace $\zeta \in L^2(\partial Q)$ of u in the sense of L^2–convergence. We have to show that $\zeta = \varphi$ a.e. on ∂Q. As in the proof of Theorem 7.2, for every $\Psi \in C^1(\bar{Q})$ we derive the following identities

$$\int_{\partial Q} d^i D_i\rho\zeta\Psi\,dS_x = \int_Q \rho a^{ij} D_i u D_j\Psi\,dx + \int_Q (c+\lambda)u\Psi\,dx$$
$$-\int_Q D_i(d^i\Psi)u\,dx - \int_Q f\Psi\,dx = \int_Q F(\Psi)\,dx$$

and similarly for u_m we have

$$\int_{\partial Q} d^i D_i\rho\varphi_m\Psi\,dS_x = \int_Q \rho a^{ij} D_i u_m D_j\Psi\,dx + \int_Q (c+\lambda)u_m\Psi\,dx$$
$$-\int_Q D_i(d^i\Psi)u_m\,dx - \int_Q f\Psi\,dx = \int_Q F_m(\Psi)\,dx.$$

Since

$$\lim_{m\to\infty} F_m(\Psi)\,dx = \int_Q F(\Psi)\,dx,$$

we have that

$$\int_{\partial Q} \varphi \Psi d^i D_i \rho \, dS_x = \int_{\partial Q} \zeta \Psi d^i D_i \rho \, dS_x$$

for every $\Psi \in C^1(\bar{Q})$ and consequently $\varphi = \zeta$ a.e. on ∂Q. The uniqueness of the solution of (7.1),(7.2) can be deduced from the following energy estimate

$$\begin{aligned} &\int_Q |D^2 u(x)|^2 \rho(x)^3 \, dx + \int_Q |Du(x)|^2 \rho(x) \, dx + \int_Q u(x)^2 \, dx \\ &\leqq C\Big[\int_Q f(x)^2 \, dx + \int_{\partial Q} \varphi(x)^2 \, dS_x\Big], \end{aligned}$$

which is valid for any $u \in \widetilde{W}^{2,2}(Q)$ satisfying (7.1),(7.2) with $\lambda \geqq \lambda_\circ$ and the proof of which is a slight modification of the proof of (7.4). We only use Lemma 7.2 in place of Theorem 7.1.

7.4. Non–local problem.

In this section we discuss the solvability of the following non–local problem in the sense Bitsadze–Samarskii (see [**BS**])

$$Lu = f(x) \text{ in } Q, \tag{7.10}$$

$$u(x) - \beta(x) u(\phi(x)) = h(x) \text{ on } \partial Q, \tag{7.11}$$

where h and β are given continuous functions on ∂Q and ϕ is a given continuous mapping of ∂Q into Q. Here L denotes the operator defined in Section 1.2 and for simplicity we assume that $b^i \equiv 0$.

This type of the boundary value problem is often refered to as the boundary value problem with the Bitsadze–Samarskii condition. We refer to the paper [**SA**] for the description of the physical aspects of non–local problems. The most significant feature of the problem (7.10),(7.11) is that the boundary condition relates values of a solution on the boundary to its values on some part of the interior of the region.

First we prove the existence of classical solution in $C^2(Q) \cap C(\bar{Q})$. Next we consider the case $h \in L^2(\partial Q)$ and we solve this problem in $\widetilde{W}^{1,2}(Q)$ using our energy estimate method. As in Section 7.1 we assume that the coefficients of L are smooth.

The uniqueness of the problem (7.10),(7.11) is a consequence of the strong maximum principle.

Lemma 7.4. *Let $|\beta(x)| \leqq 1$ on ∂Q and $c(x) \geqq 0$ in Q and suppose that either*

(a) $-1 \leqq \beta(x_\circ) < 1$ at some point $x_\circ \in \partial Q$

or

(b) $c(x_1) > 0$ at some point $x_1 \in Q$.

Then the problem (7.10),(7.11) has at most one solution in $C^2(Q) \cap C(\bar{Q})$.

PROOF. It is sufficient to show that if $f(x) \equiv 0$ on Q and $h(x) \equiv 0$ on ∂Q, then $u(x) \equiv 0$ on Q is the only solution of the problem (7.10),(7.11). It is clear that under each of the assumptions (a) or (b) any constant solution must be identically equal to 0. If $u \neq 0$, then u must be a non–constant solution and by the strong maximum principle we may assume that

$$u(x_2) = \max_{\bar{Q}} u(x) > 0 \text{ with } x_2 \in \partial Q.$$

If $\beta(x_2) = 0$ we get a contradiction. Therefore it remains to consider two cases:

$$(i)\, 0 < \beta(x_2) \leqq 1 \text{ and } (ii)\, -1 \leqq \beta(x_2) < 0.$$

In the first case $u(\phi(x_2)) = \frac{u(x_2)}{\beta(x_2)} \geq u(x_2)$, which is impossible since $\phi(x_2) \in Q$. In the second case (ii) we have $u(\phi(x_2)) = \frac{u(x_2)}{\beta(x_2)} < 0$ and by the strong maximum principle, u takes on a negative minimum at $x_3 \in \partial Q$, that is $u(x_3) = \min_{\bar{Q}} u(x) < 0$ and we may assume that $\beta(x_3) < 0$, since otherwise we get a contradiction. Hence $u(\phi(x_3)) = \frac{u(x_3)}{\beta(x_3)} > 0$. Now we distinguish two cases either: $u(x_2) \leqq |u(x_3)|$ or $u(x_2) > |u(x_3)|$. We show that both cases lead to a contradiction. Indeed, in the first case we have

$$u(x_2) \leqq |u(x_3)| = |\beta(x_3)u(\phi(x_3))| \leqq |u(\phi(x_3))| = u(\phi(x_3)),$$

which is impossible. In the second case we have

$$|u(x_3)| < u(x_2) = \beta(x_2)u(\phi(x_2)) \leqq |u(\phi(x_2))|.$$

Since both values $u(x_3)$ and $u(\phi(x_2))$ are negative, u attains its negative minimum at $\phi(x_2) \in Q$ and we arrive at a contradiction.

The existence of a solution of the problem (7.10),(7.11) can be established using the Fredholm integral equations of the second kind.

Theorem 7.4. *Suppose that the assumptions of Lemma 7.4 hold. Then the problem (7.10),(7.11) admits a unique solution* $u \in C^2(Q) \cap C(\bar{Q})$.

PROOF. We try to find a solution in the form

$$u(x) = \int_{\partial Q} \frac{dG(x,y)}{dn_y} v(y)\, dS_y - \int_Q G(x,y) f(y)\, dy, \tag{7.12}$$

where $v \in C(\partial Q)$ is to be determined, G is the Green function for the operator L and $\frac{dG}{dn_y}$ denotes the conormal derivative. The boundary condition leads to the Fredholm integral equation of the second kind

$$v(x) - \int_{\partial Q} \beta(x) \frac{dG(\phi(x),y)}{dn_y} v(v)\, dS_y = h(x) - \int_Q \beta(x) G(\phi(x),y) f(y)\, dy. \tag{7.13}$$

Since $\phi(\partial Q) \subset Q$, the kernel

$$\beta(x) \frac{dG(\phi(x),y)}{dn_y}$$

is a continuous function on $\partial Q \times Q$. By Lemma 7.4 the homogeneous equation corresponding to (7.13) has only the trivial solution. Hence by the Fredholm Alternative there exists a unique solution $v \in L^2(\partial Q)$, which by the continuity of the kernel belongs to $C(\partial Q)$. Consequently, the formula (7.12) gives a solution to the problem (7.10),(7.11).

7.5. Energy estimate for non–local problem.

In this section we consider the equation

$$Lu + \lambda u = f(x) \text{ in } Q \tag{7.14}$$

where λ is a real parameter, with the boundary condition (7.11).

We assume that $\phi : \partial Q \to Q$ is a C^1–mapping with the positive Jacobian.

Theorem 7.5. *There exist positive constants $\lambda_\circ$, C and d such that if u is a solution in $C^2(Q) \cap C(\bar{Q})$ of the problem (7.14),(7.11) for $\lambda \geqq \lambda_\circ$, then*

$$\int_Q |Du(x)|^2 r(x)\,dx + \int_Q u(x)^2 r(x)\,dx + \sup_{0<\delta\leqq d} \int_{\partial Q_\delta} u(x)^2\,dS_x$$
$$\leqq C\Big[\int_{\partial Q} h(x)^2\,dS_x + \int_Q f(x)^2\,dx\Big].$$

PROOF. We follow the proof of Theorem 3.5 (see Chapter 3). Multiplying (7.14) by

$$v(x) = \begin{cases} u(x)(\rho(x) - \delta) & \text{on } Q_\delta \\ 0 & \text{on } Q - Q_\delta, \end{cases}$$

and integrating by parts we obtain

$$\frac{1}{2}\int_{\partial Q_\delta} a^{ij} D_i\rho D_j\rho u^2\,dS_x = -\frac{1}{2}\int_{Q_\delta} D_i(a^{ij} D_j\rho)u^2\,dx + \int_{Q_\delta} a^{ij} D_i u D_j u(\rho - \delta)\,dx$$
$$+ \int_{Q_\delta} d^i D_i u u(\rho - \delta)\,dx + \int_{Q_\delta} (c + \lambda)u^2(\rho - \delta)\,dx - \int_{Q_\delta} fu(\rho - \delta)\,dx.$$

Applying Hölder's inequality we easily obtain

$$\sup_{0<\delta\leqq d} \int_{\partial Q_\delta} u^2\,dS_x \leqq C_1\Big[\int_Q |Du|^2\rho\,dx + \int_Q u^2\,dx + \lambda\int_Q u^2\rho\,dx + \int_Q f^2\,dx\Big], \tag{7.15}$$

where $d > 0$ and $C_1 > 0$ are constants. Similarly

$$\int_Q |Du|^2\rho\,dx + \lambda\int_Q u^2\rho\,dx \leqq C_2\Big[\int_Q u^2\,dx + \int_Q f^2\,dx + \int_{\partial Q} u^2\,dS_x\Big] \tag{7.16}$$

for some constant $C_2 > 0$. It follows from (7.11) that

$$\int_Q |Du|^2\rho\,dx + \lambda\int_Q u^2\rho\,dx \leqq C_3\Big[\int_Q u^2\,dx + \int_Q f^2\,dx$$
$$+ \int_{\partial Q} h^2\,dS_x + \int_{\partial Q} u(\phi(x))^2\,dS_x\Big], \tag{7.17}$$

where $C_3 > 0$. The estimates (7.15) and (7.17) yield that

$$\int_Q |Du|^2 \rho\, dx + \lambda \int_Q u^2 \rho\, dx + \sup_{0<\delta<d} u^2\, dS_x \leqq C_4 \left[\int_Q u^2\, dx + \int_Q f^2\, dx + \int_{\partial Q} h^2\, dS_x + \int_{\partial Q} u(\phi(x))^2\, dS_x \right] \tag{7.18}$$

for some $C_4 > 0$. Since a ϕ is C^1–mapping with positive Jacobian, it is obvious that

$$\int_{\partial Q} u(\phi(x))^2\, dS_x \leqq C_5 \int_{\phi(\partial Q)} u^2\, dS_x \leqq C_6 \left[\int_{Q_\phi} |Du|^2\, dx + \int_{Q_\phi} u^2\, dx \right], \tag{7.19}$$

where $C_5 > 0$ and $C_6 >$ are constants and Q_ϕ is a domain containing $\phi(\partial Q)$ with dist $(Q_\phi, \partial Q) > 0$. Consequently by the Caccioppoli inequality (see Lemma 5.1) we have

$$\int_{\partial Q} u(\phi(x))^2\, dS_x \leqq C_7 \left[\int_Q u^2\, dx + \int_Q f^2\, dx \right] \tag{7.20}$$

for some $C_7 > 0$. We now observe that

$$\int_Q u^2\, dx \leqq \frac{1}{d_1} \int_Q u^2 \rho\, dx + d \sup_{0<\delta<d} \int_{\partial Q_\delta} u^2\, dS_x, \tag{7.21}$$

where $d_1 = \inf_{Q_d} \rho(x)$. Choosing d sufficiently small and λ sufficiently large we easily derive the desired estimate from (7.16),(7.18)– (7.21).

Repeating the argument of Theorem 7.4, we deduce the following existence result.

Theorem 7.6. *Let $|\beta(x)| \leqq 1$ on ∂Q. Then there exists a positive constant $\lambda_\circ > 0$ such that for every $\lambda \geqq \lambda_\circ$ the problem (7.14),(7.11) has a unique solution in $C^2(Q) \cap C(\bar{Q})$.*

7.6. Weak solutions of the non–local problem.

The energy estimate from Section 7.6 shows that one can expect solutions of the problem (7.14),(7.11) in a weighted Sobolev space $\widetilde{W}^{1,2}(Q)$.

Guided by the results of Chapter 3 we adopt the following L^2–approach to the problem (7.14),(7.11).

Let $h \in L^2(\partial Q)$. A weak solution $u \in W^{1,2}_{\text{loc}}(Q)$ of (7.14) is a solution of the non–local problem with the boundary condition (7.11), if

$$\lim_{\delta \to 0} \int_{\partial Q} [u(x_\delta) - \beta(x) u(\phi(x)) - h(x)]^2\, dS_x = 0.$$

Here $u(\phi(x))$ is understood in the sense of trace, which is well defined since $u \in W^{1,2}_{\text{loc}}(Q)$.

We now are in a position to establish the existence result in $\widetilde{W}^{1,2}(Q)$ of the problem (7.14),(7.11).

Theorem 7.7. *Let $h \in L^2(\partial Q)$ and $|\beta(x)| \leqq 1$ on ∂Q. Then there exists a positive constant λ_o such that for $\lambda \geqq \lambda_o$ the problem (7.14),(7.11) in $\widetilde{W}^{1,2}(Q)$ admits a unique solution.*

PROOF. Let h_m be sequence in $C^1(\partial Q)$ such that $\lim_{m\to\infty} \int_{\partial Q} [h_m - h]^2\, dS_x = 0$. Let λ_o be a constant from Theorem 7.5 and assume that $\lambda \geqq \lambda_o$. For each $m \geqq 1$ Theorem 7.4 guarantees the existence of a unique solution $u_m \in C^2(Q) \cap C(\bar{Q})$ to the problem (7.14),(7.11). Moreover, we have for each m

$$\int_Q |Du_m|^2 \rho\, dx + \int_Q u_m^2\, dx \leqq C\left[\int_Q f^2\, dx + \int_{\partial Q} h_m^2\, dS_x\right],$$

where $C > 0$ is a constant independent of m. Since the sequence u_m is bounded in $\widetilde{W}^{1,2}(Q)$, there exists a subsequence, which we relabel as u_m, converging weakly in $\widetilde{W}^{1,2}(Q)$ to a function u. By Theorem 1.3, $\widetilde{W}^{1,2}(Q)$ is compactly imbedded in $L^2(Q)$ and therefore we may assume that u_m converges to u in $L^2(Q)$. It is obvious that u is a weak solution of (7.14). By Theorem 3.4, u has a trace $\zeta \in L^2(\partial Q)$ in the sense of L^2-convergence, that is

$$\lim_{\delta\to 0} \int_{\partial Q} [u(x_\delta) - \zeta(x)]^2\, dS_x = 0.$$

To complete the proof we show that $\zeta(x) = \beta(x)u(\phi(x)) + h(x)$ a.e. on ∂Q. Let $\Phi \in C^1(\bar{Q})$. It is easy to see that $\Phi(x)\rho(x)$ is a legitimate test function and integrating by parts we obtain

(7.22)

$$\int_{\partial Q} \zeta \Phi a^{ij} D_i\rho D_j\rho\, dS_x = -\int_Q a^{ij} D_i(a^{ij} D_j\rho\Phi)u\, dx + \int_Q a^{ij} D_i u D_j \Phi\rho\, dx$$
$$+ \int_Q d^i D_i u \Phi\rho\, dx + \int_Q (c+\lambda)u\Phi\rho\, dx - \int_Q f\Phi u\rho\, dx \equiv F(u).$$

Similarly

$$\int_{\partial Q} [h_m(x) + \beta(x)u_m(\phi(x))]\Phi a^{ij} D_i\rho D_j\rho\, dS_x \equiv F(u_m). \tag{7.23}$$

Applying the estimate (7.19) and the obvious analogue of the energy estimate to $u_p - u_q$ we obtain

$$\int_{\partial Q} [u_p(\phi(x)) - u_q(\phi(x))]^2\, dS_x \leqq C\left[\int_{Q_\phi} |Du_p - Du_q|^2\, dx + \int_Q [u_p - u_q]^2\, dx\right]$$
$$\leqq \frac{C}{k}\int_Q |Du_p - Du_q|^2 \rho\, dx + C\int_Q [u_p - u_q]^2\, dx \leqq \tilde{C}\int_{\partial Q} |h_p - h_q|^2\, dS_x,$$

where $k = \inf_{\bar{Q}_\phi} \rho(x)$ and $\tilde{C}$ is a constant independent of p and q. Hence by the continuity of weak solutions on Q we may assume that $u_m(\phi(x)) - u(\phi(x)) \to 0$ as $m \to \infty$ in $L^2(\partial Q)$. Combining this with the fact that $F(u_m) \to F(u)$ as $m \to \infty$ we deduce from (7.22),(7.23) that $\zeta(x) = \beta(x)u(\phi(x)) + h(x)$ a.e. on ∂Q and this completes the proof.

CHAPTER 8

DOMAINS WITH $C^{1,\alpha}$–BOUNDARY

In this chapter we show that the assumption that the boundary ∂Q of Q is of class C^2 can be replaced by $C^{1,\alpha}$–regularity. This was first observed by Petrushko (see [**PE**]). In his approach he used, as distance–like function ρ, a solution of the problem

$$\Delta\rho = -1 \text{ in } Q,\ \rho(x) = 1 \text{ on } \partial Q,$$

with ∂Q of class $C^{1,\alpha}$. It is well known that $\rho \in C^{1,\alpha}(\bar{Q}) \cap C^2(Q)$ and there exists a constant $\gamma_1 > 0$ such that

$$\gamma_1^{-1} r(x) \leqq \rho(x) \leqq \gamma_1 r(x).$$

In addition, there exists a constant $C = C(\alpha)$ such that

$$|D^2\rho(x)| \leqq \frac{C(\alpha')}{r(x)^{1-\alpha'}}$$

for all $0 < \alpha' < \alpha$. By a boundary point lemma there exist numbers $r_\circ > 0$ and $\gamma > 0$ such that $|D\rho(x)| > \gamma$ for all $x \in \bar{Q} \times \{r(x) \leqq r_\circ\}$. One can prove that with this choice of ρ, Lemma 2.1 continues to hold, but the mapping x_δ can be defined locally. Therefore the proof of the theorem giving the existence of traces of solutions in $\widetilde{W}^{1,2}(Q)$ requires some modifications. The aim of this chapter is to show the distance–like function ρ defined in Chapter 1 can be replaced by regularized distance which has the same properties as the function ρ used by Petrushko. The regularized distance was constructed by Gilbarg–Hörmander (see [**GH**]) and used to obtain the intermediate Schauder estimates. In our construction of the regularized distance, presented in this chapter, we closely follow the paper by Lieberman (see [**L1**]). We also briefly discuss a recent development, due to Lieberman (see [**L2**]), on the use of the regularized distance in L^2–approach for the Dirichlet problem in domains with $C^{1,\mathrm{Dini}}$–boundaries.

8.1. Regularized distance.

To weaken the assumption on the regularity of the boundary we use the concept of the regularized distance. We follow the ideas of the paper [**L1**].

Let Q be an open and bounded subset of R_n. We define the signed distance to the boundary ∂Q by

$$d(x) = \begin{cases} \text{dist }(x, \partial Q) & \text{for } x \in Q \\ -\text{ dist }(x, \partial Q) & \text{for } x \notin Q. \end{cases}$$

A function ρ is said to be a regularized distance for Q if $\rho \in C^2(R_n - \partial Q) \cap C^{0,1}(R_n)$ and if the ratios $\frac{\rho(x)}{d(x)}$ and $\frac{d(x)}{\rho(x)}$ are positive and uniformly bounded for all $x \in R_n - \partial Q$. The construction of the regularized distance is based on the following lemma.

Lemma 8.1. *Let Q be an open bounded subset of R_n and suppose that there is a Lipschitz function g for which the ratios $\frac{d}{g}$ and $\frac{g}{d}$ are uniformly bounded and positive in $R_n - \partial Q$. Let $K > 0$ be constant such that $|g(x) - g(y)| \leqq \frac{K}{2}|x - y|$ for all x and* *y in R_n, let ϕ be a nonnegative function in $C^2(R_n)$ with supp $\phi \subset B(0,1)$ such that $\int_{R_n} \phi(x)\,dx = 1$ and define*

$$G(x,\tau) = \int_{\{|z|\leqq 1\}} g(x - \frac{\tau}{K}z)\phi(z)\,dz.$$

Then a regularized distance is given by the equation

$$\rho(x) = G(x, \rho(x)). \tag{8.1}$$

PROOF. We commence by investigating the properties of G. For $\tau \neq 0$, we can write

$$G(x,\tau) = (-\frac{K}{\tau})^n \int_{R_n} g(z)\phi(\frac{K(x-z)}{\tau})\,dz; \tag{8.2}$$

it is clear from this formula that $G \in C^2(R_{n+1} - \{x, 0\})$, since g is continuous and $\operatorname{supp}\phi \subset B(0,1)$. Moreover

$$G(x,\tau_1) - G(x,\tau_2) = \int_{\{|z|\leqq 1\}} [g(x - \frac{\tau_1}{K}z) - g(x - \frac{\tau_2}{K}z)]\phi(z)\,dz,$$

so the choice of K implies that

$$|G(x,\tau_1) - G(x,\tau_2)| \leqq \int_{\{|z|\leqq 1\}} \frac{1}{2}K|\tau_1 - \tau_2|(\frac{|z|}{K})\phi(z)\,dz \leqq \frac{1}{2}|\tau_1 - \tau_2|. \tag{8.3}$$

Similarly

$$|G(x_1,\tau) - G(x_2,\tau)| \leqq \frac{K}{2}|x_1 - x_2|. \tag{8.4}$$

We now consider (8.1). It follows (8.3) and the Banach fixed point theorem that (8.1) has a unique continuous solution ρ. Another application of (8.3) yields

$$G(x,0) - \frac{1}{2}|\rho(x)| \leqq \rho(x) \leqq G(x,0) + \frac{1}{2}|\rho(x)|.$$

Since $G(x,0) = g(x)$, we see that $\frac{2}{3} \leqq \frac{\rho}{g} \leqq 2$ and hence $\frac{\rho}{d}$ and $\frac{d}{\rho}$ are positive and uniformly bounded. Combining (8.3), (8.4) and the equation

$$\rho(x) - \rho(y) = \Big(G(x,\rho(x)) - G(x,\rho(y))\Big) + \Big(G(x,\rho(y)) - G(y,\rho(y))\Big)$$

yields

$$|\rho(x) - \rho(y)| \leqq K|x - y|, \tag{8.5}$$

so ρ is Lipschitz. The fact that ρ is C^2 is a consequence of the implicit function theorem and therefore ρ is the regularized distance.

It is well known that for every domain Q, $d(x, \partial Q)$ is a Lipschitz function in x. Consequently applying Lemma 8.1 with $g = d$ we obtain

Corollary 8.1. *Every domain has a regularized distance.*

In the next theorem we show that the regularity of derivatives of a regularized distance ρ depends on the regularity of g.

Theorem 8.2. *Let Q and g be as in Lemma 8.1 and suppose that $g \in C^1(R_n)$. Then Q has a C^1–regularized distance. Let ζ be a continuous, increasing function with $\zeta(0) = 0$. If*

$$|Dg(x) - Dg(y)| \leqq \zeta(|x-y|) \text{ for all } x, y \in R_n,$$

and if K and ϕ are as in Lemma 8.1, then

$$|D\rho(x) - D\rho(y)| \leqq 8\zeta(|x-y|) \text{ for all } x, y \in R_n, \tag{8.6}$$

$$|D_{ij}\rho(x)| \leqq 2(4K_1 + 3n + 1)\frac{K}{\rho(x)}\zeta\left(\frac{|\rho(x)|}{K}\right) \text{ for all } x \in R_n - \partial Q, \tag{8.7}$$

where $K_1 = \int_{|x|\leqq 1} |D\phi(x)|\, dx$. Moreover, there exist positive constants c_1 and c_2 such that

$$|D\rho(x)| \geq c_1 \text{ whenever } 0 < |\rho(x)| \leqq c_2. \tag{8.8}$$

PROOF. It is clear that G is a C^1–function and consequently the implicit function theorem implies that ρ is a C^1–function. To proceed further we introduce notations

$$G_i(x) = \frac{\partial G}{\partial x_i}(x, \rho(x)),\ G_{n+1}(y) = \frac{\partial G}{\partial \tau}(y, \rho(y)),$$

where τ is identified with x_{n+1}. We also put $G'(x) = \big(G_1(x), ..., G_n(x)\big)$. Differentiating (8.1) we obtain

$$\begin{aligned} D\rho(x) - D\rho(y) = \big(G'(x) - G'(y)\big) + D\rho(y)\big(G_{n+1}(x) - G_{n+1}(y)\big) \\ + G_{n+1}(x)\big(D\rho(x) - D\rho(y)\big) \end{aligned}$$

and combining this with (8.3) and (8.5) we have

$$|D\rho(x) - D\rho(y)| \leqq 2|G'(x) - G'(y)| + 2K|G_{n+1}(x) - G_{n+1}(y)|. \tag{8.9}$$

We now observe that

$$\begin{aligned} G'(x) &= \int_{\{|z|\leqq 1\}} Dg\big(x - \frac{\rho(x)}{K}z\big)\phi(z)\, dz \\ G_{n+1}(x) &= -\frac{1}{K}\int_{\{|z|\leqq 1\}} zDg\big(x - \frac{\rho(x)}{K}z\big)\phi(z)\, dz. \end{aligned} \tag{8.10}$$

Therefore

$$|G'(x) - G'(y)| \leqq \int_{\{|z|\leqq 1\}} |Dg(x - \frac{\rho(x)}{K}z) - Dg(y - \frac{\rho(x)}{K})|\phi(z)\,dz$$
$$+ \int_{\{|z|\leqq 1\}} |Dg(y - \frac{\rho(x)}{K}z) - Dg(y - \frac{\rho(y)}{K}z)|\phi(z)\,dz$$
$$\leqq \int_{\{|z|\leqq 1\}} \left[\zeta(|x-y|) + \zeta(\frac{|\rho(x) - \rho(y)|}{K})\right]\phi(z)\,dz \leqq 2\zeta(|x-y|).$$

Similarly

$$|G_{n+1}(x) - G_{n+1}(y)| \leqq \frac{2\zeta(|x-y|)}{K}.$$

Inserting the last two estimates into (8.9), we obtain (8.6). To derive (8.7) we differentiate (8.2) twice to obtain

(8.12)
$$D_{ij}\rho(x) = \frac{1}{1 - G_{n+1}(x)}\Big[G_{ij}(x) + G_{i,n+1}(x)D_j\rho(x) + G_{j,n+1}(x)D_i\rho(x)$$
$$+ G_{n+1,n+1}(x)D_i\rho(x)D_j\rho(x)\Big].$$

We see that to derive (8.7) we must estimate the second derivatives of G. It follows from (8.10) that

(8.13)
$$G_{ij}(x) = -\frac{K}{\rho(x)}\int_{\{|z|<1\}} D_i g(x - \frac{\rho(x)}{K}z)D_j\phi(z)\,dz$$
$$G_{i,n+1}(x) = -\frac{1}{\rho(x)}\int_{\{|z|<1\}} D_i g(x - \frac{\rho(x)}{K}z)D_j\big(z_j\phi(z)\big)\,dz$$
$$G_{n+1,n+1}(z) = -\frac{1}{K\rho(x)}\int_{\{|z|<1\}} D_k g(x - \frac{\rho(x)}{K}z)D_j\big(z_k z_j\phi(z)\big)\,dz.$$

This can be achieved by replacing $\rho(x)$ in equations (8.10) by τ, making the substitution $y = x - \frac{\tau}{K}z$, differentiating, then converting back to the variable z. Since $\int_{\{|z|<1\}} D_j\phi(z)\,dz = 0$, we have

$$G_{ij}(x) = -\frac{K}{\rho(x)}\int_{\{|z|<1\}}\left[D_i g(x - \frac{\rho(x)}{K}z) - D_i g(x)\right]D_j\phi(z)\,dz.$$

We also have analogous expressions for $G_{i,n+1}(x)$ and $G_{n+1,n+1}(x)$. Now the estimate

$$|D_i g(x - \frac{\rho(x)}{K}z) - D_i g(x)| \leqq \zeta(\frac{|\rho(x)|}{K})$$

implies that

$$|G_{ij}(x)| \leqq K_1 K \frac{\zeta(\frac{|\rho(x)|}{K})}{|\rho(x)|},$$

$$|G_{i,n+1}(x)| \leqq (n + K_1)\frac{\zeta(\frac{|\rho(x)|}{K})}{|\rho(x)|}$$

and

$$|G_{n+1,n+1}(x)| \leqq (n + 1 + K_1)\frac{\zeta(\frac{\rho(x)|}{K})}{|\rho(x)|K}.$$

These inequalities combined with (8.3), (8.5) and (8.12) give (8.7).
To prove (8.8) we show that $|D\rho(x)| > 0$ on ∂Q. Then the continuity of $D\rho$ yields (8.8). If $x \in \partial Q$, then

$$|D\rho(x)| = |Dg(x)|[1 + Dg(x) \int_{\{|z|<1\}} z\phi(z)\, dz]^{-1} \geqq \frac{2|Dg(x)|}{2 + K},$$

since $|Dg(x)| \leqq \frac{K}{2}$. Thus $|\rho(x)| > 0$ on ∂Q, if $|Dg(x)| > 0$ on ∂Q, so let $x \in \partial Q$. Suppose first that Q satisfies an interior sphere condition at x with center x_o and let $M > 0$ be constant such that $g(y) \geq Md(y)$ for all $y \in R_n$. If ω is a vector from x to x_o, it is obvious that $d(x + t\omega) = t|\omega|$ for $0 \leqq t \leqq 1$, and hence

$$g(x + t\omega) - g(x) = g(x + t\omega) \geqq Mt|\omega|$$

for $0 \leqq t \leqq 1$. This implies that $\omega Dg(x) \geqq M|\omega|$ and consequently $|Dg(x)| \geqq M$.
To complete the proof, we need only show that the set of points of ∂Q at which Q satisfies an interior sphere condition is dense in ∂Q. Let $x \in \partial Q$ and let $\delta > 0$ be arbitrary and choose $x_1 \in Q$ and $x_2 \in \partial Q$ such that

$$|x - x_1| < \frac{\delta}{2}, \; |x_1 - x_2| = d(x_1).$$

Since $d(x_1) < \frac{\delta}{2}$, it follows that $|x - x_2| < \delta$. Consequently Q satisfies an interior sphere condition (with center x_1 and radius $d(x_1)$) at x_2. Therefore the set of points of ∂Q at which Q satisfies an interior sphere condition is dense in ∂Q.

We now show that local properties of ρ are determined by local properties of ∂Q. To explore these local properties we need a local regularized distance.

Let $x_o \in \partial Q$, and define $\Omega_r = B_r \cap Q$, where $B_r = B(x_o, r)$ for some $r > 0$. If there is a constant $r > 0$ and a function $\rho \in C^2(B_r - \partial Q)$ which is Lipschitz continuous on $\bar{B}_r$ and such that the ratios $\frac{d}{\rho}$ and $\frac{\rho}{d}$ are bounded and positive on $B_r - \partial Q$, we call ρ a local regularized distance at x_o.

We also need the definition of a local representation of ∂Q at x_o (see Section 1.2). If there are positive constants δ and A an orthonormal coordinate system $y = (y', y_n) = (y_1, ..., y_{n-1}, y_n)$ with the origin at x_o, and a function f such that $\Omega_{4\delta} = \{y \in B_{4\delta} : y_n > f(y')\}$ and

$$|f(y_1') - f(y_2')| \leqq A|y_1' - y_2'| \text{ for all } y_1', y_2' \text{ in } B(x_o, 4\delta),$$

we call f a local representation for ∂Q at x_o.

For a fixed $x_o \in \partial Q$ at which a local representation f exists, it is easy to verify that $g(y) = y_n - f(y')$ obeys the hypotheses of Lemma 8.1 with R_n replaced by $B_{2\delta}$. Obviously g is Lipschitz in $B_{2\delta}$ with

$$|g(x) - g(y)| \leqq \max\{1, A\}|x - y|$$

for x, y in $B_{2\delta}$ and the ratios $\frac{d}{g}$ and $\frac{g}{d}$ are positive in $B_{2\delta}$ with

$$|d(x)| \leqq |g(x)| \leqq (A^2 + 1)^{\frac{1}{2}}$$

for $x \in B_{2\delta}$. If we choose $K = 2\max\{1, A\}$, the construction of ρ in Lemma 8.1 can be carried out for $x \in B_\delta$. To see this, we observe first that $G(x, \tau)$ is defined for $x \in B_\delta$ and $\frac{|\tau|}{K} \leqq \delta$. Since $f(0) = 0$, it follows from (8.4) that $|G(x, \tau)| \leqq \frac{K|x|}{2} + \frac{\tau}{2}$. Hence for each $x \in B_\delta$, $G(x, .)$ maps the interval $[-K\delta, K\delta]$ into itself and consequently the construction from Lemma 8.1 can be carried out in this case. It is clear that $\frac{\rho}{K} < \delta$. To show that ρ satifies (8.8) we observe that $g(x + \epsilon w) - g(x) = \epsilon$ for a unit vector w in the y_n–direction, so $G_n(x) = 0$ on B_δ. In fact we have $|D\rho(x)| > \frac{1}{2}$ in B_δ.

Repeating the argument of Theorem 8.1 we obtain

Theorem 8.2. *Let $x_o \in \partial Q$. If ∂Q has a C^1 local representation f at x_o then Q has a C^1 local regularized distance ρ at x_o. Moreover if ζ is an increasing function such that $\zeta(0) = 0$ and if*

$$|Df(x') - Df(y')| \leqq \zeta(|x' - y'|) \text{ for } x', y' \in B_{4\delta},$$

then there is a positive constant C_1, determined only by A, δ and n, such that

$$|D\rho(x) - D\rho(y)| \leqq 10\zeta(|x - y|) \text{ for } x, y \in B_\delta$$

$$|D^2\rho(x)| \leqq C_1 \frac{\zeta(\frac{\rho(x)}{2}\max\{1, A\})}{|\rho(x)|} \text{ for } x \in B_\delta.$$

One can show that both constructions of regularized distance are equivalent. As an easy application of partition of unity we obtain:

Theorem 8.3. *Let Q be an open set in R_n. If there is a function $g \in C^1(R_n)$ such that the hypotheses of Lemma 8.1 are satisfied, then there is a local representation for ∂Q at every point of ∂Q. If there is a local representation for ∂Q at each point of ∂Q and if the constants A and δ of the definition of local representation can be chosen independently of the particular point x_o, then there is a function $g \in C^1$ satisfying the hypotheses of Lemma 8.1.*

Combination of Theorems 8.2 and 8.3 yields the following result.

Theorem 8.4. *Suppose that the boundary ∂Q is of class $C^{1,\alpha}$ for $0 < \alpha < 1$. Then there is a regularized distance ρ in $C^2(Q) \cap C^1(\bar{Q})$ such that $|D\rho(x)| > c_1$ for $\rho(x) < c_2$ for some constants $c_1 > 0$ and $c_2 > 0$. Moreover, $\frac{\rho}{d}$ is bounded above and below on Q and*

$$|D^2\rho(x)| \leqq C_3\rho(x)^{\alpha-1} \text{ on } Q$$

for some constant $C_3 > 0$.

8.2. Analogue of Lemma 2.1.

The aim of this section is to show that Lemma 2.1 remains true for ∂Q of class $C^{1,\alpha}$. To this end we repeat a slightly modified construction from Lemma 6.2.

Let ρ be a regularized distance from Lemma 8.2 with $\zeta(t) = t^\alpha$. Let us choose a number $\delta_\circ > 0$ so that the the domain $Q - Q \cap \{r(x) \geqq \delta_\circ\}$ contains the level surfaces of $\rho(x)$: $\rho(x) = \delta$, $0 < \delta \leqq \delta_\circ$, $x \in Q$. We also assume that $\delta_\circ$ is sufficiently small so that the subset $Q_\delta = Q \cap \{x \in Q;\ \rho(x) > \delta\}$, $0 < \delta \leqq \delta_\circ$, is a domain with the boundary ∂Q_δ of class C^2 and the normal at every point of ∂Q intersects ∂Q_δ for all $\delta \in (0, \delta_\circ]$. We now observe that $\partial Q_\circ = \partial Q$ and ∂Q_δ is of class $C^{1,\alpha}$ for all $\delta \in [0, \delta_\circ]$. Let us fix a point $x^\circ \in \partial Q$. We denote by $x^\circ_{\delta_\circ}$ the closest point of $\partial Q_{\delta_\circ}$ to the point x°. This point lies on the intersection of the normal to ∂Q at this point with $\partial Q_{\delta_\circ}$. Let $0, y_1, \ldots, y_n$ be an orthogonal system of coordinates such that $x^\circ_{\delta_\circ}$ is the origin of coordinates, and the outward normal to ∂Q at x° has the direction $0, y_n$, This system will be called a local coordinate system. The coordinates of a point x in the local coordinate system will be denoted by $(y_1, \ldots, y_n) = (y', y_n)$. The coordinates of x° are $(0, \ldots, 0, y_n^\circ)$. The function $\rho(x)$ in the local coordinate system is denoted by $\tilde{\rho}(y)$.

We now consider the function $R : [0, \delta_\circ] \times Q \to R$ given by $R(\delta, y', y_n) = \tilde{\rho}(y) - \delta$. For fixed $\delta \in (0, \delta_\circ]$ the zero level surface of R coincides with ∂Q_δ. Since

$$\frac{\partial R}{\partial y_n}(0, 0, y_n^\circ) = \frac{\partial \rho(x^\circ)}{\partial \nu} < 0,$$

by the implicit function theorem there exist positive numbers r_1, $\delta_1 < \frac{\delta_\circ}{2}$ and h such that for $\delta \in [0, \delta_1]$ the connected piece Γ_δ of Q_δ in the intersection $\partial Q_\delta \cap \{y;\ |y'| < r,\ y_n^\circ - h < y_n < y_n^\circ + h\}$ is given by the equation $y_n = \varphi(\delta, y')$, where $\varphi \in C^1([0, \delta_1] \times \{|y'| < r_1\})$. We also choose r_1 sufficiently small so that the hyperplane $y_n = 0$ does not intersect the part of Γ_δ of ∂Q_δ for any $\delta \in (0, \frac{\delta_\circ}{2}]$, the angle between the normal to ∂Q at x° and the normal to the surface $\Gamma_\circ(x^\circ) \subset \partial Q$ at any point does not exceed $\frac{\pi}{8}$, where $\Gamma_\circ(x^\circ) = \partial Q \cap \{y;\ |y'| < r_1,\ y_n^\circ - h < y_n < y_n^\circ + h\}$, and for any $\delta \in (0, \delta_1]$ the cylinder $\Omega^h = \{y;\ |y'| < r_1,\ 0 < y_n < \varphi(0, y') + h\}$ does not contain points of ∂Q_δ other than the points of the surface Γ_δ described by the equation $y_n = \varphi(\delta, y')$ for $|y'| < r_1$.

Let $0 < \delta \leqq \delta_1$ and let $\Omega_\delta = \{y;\ |y'| < r_1,\ 0 < y_n < \varphi(\delta, y')\}$ and $\Omega = \Omega^h \cap Q$. We construct a mapping $A_\delta : \Omega_\delta \to \Omega$ as follows: a point $x \in \Omega_\delta$ with local coordinates (y', y_n) is mapped into the point $A_\delta(x) \in \Omega$ with local coordinates

$$(y', \frac{y_n}{\varphi(\delta, y')}\varphi(0, y'))$$

It is clear that the surface Γ_δ is mapped into $\Gamma_\circ \equiv \Gamma_\circ(x^\circ)$. The inverse mapping A_δ^{-1} is defined similarly. A point $x \in \Omega$ with local coordinates (y', y_n) is mapped into the point $A_\delta^{-1}(x) \in \Omega_\delta$ with local coordinates

$$(y', \frac{y_n}{\varphi(0, y')}\varphi(\delta, y')).$$

It follows from the properties of φ that the mappings A_δ and A_δ^{-1} are in C^1 for all $\delta \in (0, \delta_1]$. The mapping of $\Gamma_\circ(x^\circ)$ into Γ_δ will be denoted by $x_{\delta x^\circ}(x)$, and the mapping of Γ_δ into $\Gamma_\circ(x^\circ)$, inverse to it, will be denoted by $x_{\delta x^\circ}^{-1}(x)$.

To show that Lemma 2.1 remains true for ∂Q of class $C^{1,\alpha}$ we choose $\delta \in (0, \delta_\circ]$ and let x_δ be any point of the surface Γ_δ. We set

$$\frac{dS_\delta}{dS_\circ}(x_\delta) = \lim_{\epsilon \to 0} \frac{|A_\epsilon|}{|B_\epsilon|},$$

where $A_\epsilon = \Gamma_\delta \cap \{|x - x_\delta| < \epsilon\}$ and $B_\epsilon = \{x \in \partial Q;\ x = x_{\delta x^\circ}^{-1}(\bar{x}_\delta),\ \bar{x}_\delta \in \Gamma_\delta \cap \{|x - x_\delta| < \epsilon\}$. As a direct consequence of properties of the function φ we obtain

$$\gamma_\circ^{-1} \leqq \frac{dS_\delta}{dS_\circ} \leqq \gamma_\circ \text{ and } \frac{dS_\delta}{dS_\circ}(x_\delta) \to 1 \text{ as } \delta \to 0$$

for some constant $\gamma_\circ > 0$.

8.3. Energy estimate, the Dirichlet problem and traces.

Since the mapping $x_{\delta x^\circ}$ is defined locally we adopt the following approach to the Dirichlet problem.

Let $\varphi \in L^2(\partial Q)$. A weak solution $u \in W^{1,2}_{\text{loc}}(Q)$ of (1.1) is a solution of the Dirichlet problem with the boundary condition $u(x) = \varphi(x)$ on ∂Q, if for each point $x^\circ \in \partial Q$ there exists a neighbourhood $\mathcal{U}(x^\circ) \subset \Gamma_\circ(x^\circ)$ such that

$$\lim_{\delta \to 0} \int_{\mathcal{U}(x_\circ)} \left(u(x_{\delta x^\circ}) - \varphi(x)\right)^2 dS_x = 0. \tag{8.14}$$

We now observe that the the energy estimate (3.31) for the problem (3.30) remains true (see Theorem 3.5). Using the approximation argument of Theorems 2.5 and 3.6 we easily deduce the existence result.

Theorem 8.4. *For each $\lambda \geqq \lambda_\circ$ the problem (3.30) has a unique solution in $\widetilde{W}^{1,2}(Q)$.*

Obviously the boundary condition here is understood in the sense of (8.14).

It is also clear that Theorem 3.1 of Chapter 3 continues to hold, that is, the conditions (I) and (II) are equivalent.

We now show that each of the conditions (I) and (II) of Theorem 3.1 implies the existence of a trace in $L^2(\partial Q)$ on ∂Q. In Theorem 8.5 we follow the argument from paper by Petrushko (see [PE] and also ideas developed in Chapters 2 and 3).

Theorem 8.5. *Let $u \in W^{1,2}_{\text{loc}}(Q)$ be a solution of (1.1) such that one of the conditions (I) or (II) holds. Then there is a function φ belonging to $L^2(\partial Q)$ such that for each $x^\circ \in \partial Q$ there is a neighbourhood $\mathcal{U}(x^\circ) \subset \Gamma_\circ(x^\circ)$ such that (8.14) holds.*

PROOF. First we suppose that the domain Q with the boundary of class $C^{1,\alpha}$ is strictly star–shaped with respect to one of its points. In this case we may assume that $Q = \{x;\ |x| < F(x)\}$ and $\partial Q = \{x;\ |x| = F(x)\}$, where $F(x)$ is a positive homogeneous function of degree 0. Together with subdomains Q_δ we consider subdomains $Q^\delta =$

$\{x; |x| < (1-\delta)F(x)\}$ with boundaries $\partial Q^\delta = \{x;\, |x| = (1-\delta)F(x)\}$. It is obvious that the condition

$$\sup_{\{0<\delta<\delta_0\}} \int_{\partial Q^\delta} u(x)^2 \, dS_x < \infty,$$

for some small δ_0, is equivalent to the conditions (I) and (II). Consequently it is not difficult to show that there exists $\varphi \in L^2(\partial Q)$ such that

$$\lim_{\delta \to 0} \int_{\partial Q} u((1-\delta)x) g(x) \, dS_x = \int_{\partial Q} \varphi(x) g(x) \, dS_x$$

for all $g \in L^2(\partial Q)$. Repeating the argument from the proof of Theorem 3.2, we show that

$$\lim_{\delta \to 0} \int_{\partial Q} [u((1-\delta)x) - \varphi(x)]^2 \, dS_x = 0. \tag{8.15}$$

Finally we consider a general case where ∂Q is of class $C^{1,\alpha}$. We choose an arbitrary point $x^\circ \in \partial Q$. There exists a sufficiently small number $R(x^\circ)$ and a star-shaped domain $Q(x^\circ) \subset Q$ with $\partial Q(x^\circ)$ of class $C^{1,\alpha}$ such that

$$\partial Q(x^\circ) = \Gamma(x^\circ) \cup \Gamma^1(x^\circ),$$

where $\Gamma(x^\circ) = \partial Q \cap B(x^\circ, R(x^\circ))$ and $\Gamma^1(x^\circ) \subset Q$. Choosing a finite subcovering of the covering $\{\Gamma(x^\circ),\, x^\circ \in \partial Q\}$ of the boundary ∂Q, we get a finite number of domains $Q_1(x_1^\circ), ..., Q_{N_1}(x_{N_1}^\circ)$ such that

$$\bigcup_{i=1}^{N_1} \Gamma(x_i^\circ) = \partial Q.$$

From now on we denote these domains by $Q_1, ..., Q_{N_1}$. For every $1 \leqq i \leqq N_1$ we denote by $r_i(x)$ the distance from a point $x \in Q_i$ to ∂Q_i. Since $r_i(x) \leqq r(x)$ for all $x \in Q_i$, we have

$$\int_{Q_i} |Du(x)|^2 r_i(x) \, dx \leqq \int_Q |Du(x)|^2 r(x) \, dx < \infty.$$

By the previous part of the proof there exists a function $\varphi_i \in L^2(\partial Q_i)$ such that u converges to φ_i in the sense of (8.15). We now show that φ_i is also a limit of u on $\Gamma(x_i^\circ)$ in the sense of (8.14), that is for any $x_1 \in \Gamma(x_i^\circ)$ there exists a neighbourhood of $\mathcal{U}^i(x_1) \subset \Gamma(x_i^\circ)$, such that

$$\lim_{\delta \to 0} \int_{\mathcal{U}^i(x_1)} \left(u(x_{\delta x_1}(x)) - \varphi_i(x)\right)^2 dS_x = 0. \tag{8.16}$$

This will complete the proof, because if $\Gamma(x_i^\circ) \cap \Gamma(x_j^\circ) \neq \emptyset$, then $\varphi_i(x) = \varphi_j(x)$ for all $x \in \Gamma(x_i^\circ) \cap \Gamma(x_j^\circ)$. In other words, there exists a function $\varphi \in L^2(\partial Q)$ such that (8.14) holds.

To demonstrate (8.14) we choose an arbitrary point $x_1 \in \Gamma(x_i^\circ)$ and construct a local system of coordinates (y', y_n), defined in Section 8.2. We now select a number $r > 0$ sufficiently small so that

$$B_1(x_1) = \{y : |y'| < r, 0 < y_n < \varphi(0, y') + h\} \cap \partial Q \subset \Gamma(x_i^\circ) \cap \Gamma_\circ(x_1).$$

Let $\delta^1 \leqq \frac{\delta_0}{2}$ be such that

$$\Gamma_{i,\delta} = \partial Q_\delta \cap \{|y'| < r, 0 < y_n < \varphi(0, y')\} \subset Q$$

for all $0 < \delta \leqq \delta^1$. We put

$$\begin{aligned}
\mathcal{U}_1(x_1) &= \{y;\ |y'| < r,\ \varphi(\delta^1, y') < y_n < \varphi(0, y')\},\\
B_2(x_1) &= \{y; |y'| < \frac{3}{4}r,\ 0 < y_n < \varphi(0, y') + h\} \cap \partial Q.\\
\mathcal{U}_2(x_1) &= \{y;\ |y'| < \frac{3}{4}r,\ \varphi(\frac{3}{4}\delta^1, y') < y_n < \varphi(0, y')\},\\
B_3(x_1) &= \{y;\ |y'| < \frac{r}{2},\ 0 < y_n < \varphi(0, y') + h\} \cap \partial Q,\\
\mathcal{U}_3(x_1) &= \{y;\ |y'| < \frac{r}{2},\ \varphi(\frac{\delta^1}{2}, y') < y_n < \varphi(0, y')\},\\
\Gamma'_{i\delta} &= \partial Q \cap \{y;\ |y'| < \frac{3}{4}r,\ 0 < y_n < \varphi(0, y')\}, \delta \in (0, \delta^1].
\end{aligned}$$

Let the function $\zeta \in C^\infty(R_n)$ be such that $\zeta(x) = 1$ for $x \in \overline{\mathcal{U}_3(x_1)}$, $\zeta(x) = 0$ for $x \in \bar{Q} - \overline{\mathcal{U}_2(x_1)}$, and define $v(x) = u(x)\zeta(x)$. Then v is a solution in $W^{1,2}_{\mathrm{loc}}(Q)$ of the equation $Lv = f_1 - D_i g^i$, where

$$f_1 = f\zeta - a^{ij} D_i u D_j \zeta - b^i u D_i \zeta + d^i u D_i \zeta, \quad g^i = a^{ji} u D_j \zeta.$$

By the previous part of the proof

$$\lim_{\delta \to 0} \int_{\partial Q_i} \left(v((1-\delta)x) - \zeta(x)\varphi_i(x)\right)^2 dS_x = 0.$$

Since v satisfies each of the conditions (I) and (II), therefore

$$\sup_{0<\delta<\delta^1} \int_{\Gamma^1_{i\delta}} v(x)^2\, dS_x < \infty.$$

Hence, by a result of Section 8.2, we have

$$\sup_{0<\delta<\delta^1} \int_{B_2(x_1)} v(x_{\delta x_1}(x))^2\, dS_x < \infty,$$

that is the collection of functions $\{v(x_{\delta x_1}(x); 0 < \delta \leqq \delta^1\}$ is weakly compact in $L^2(B_2(x_1))$. As in Theorem 3.3 we prove that there exists a function $\tilde{\varphi} \in L^2(B_2(x_1))$ such that

$$\lim_{\delta \to 0} \int_{B_2(x_1)} v(x_{\delta x_1}(x)) g(x)\, dS_x = \int_{B_2(x_1)} \tilde{\varphi}(x) g(x)\, dS_x$$

for each $g \in L^2(B_2(x_1))$. We now show that $\tilde{\varphi}(x) = \zeta(x)\varphi_i(x)$ on $B_2(x_1)$. To prove this it suffices to verify that

$$\lim_{\delta \to 0} \int_{B_2(x_1)} \big(v(x_{\delta x_1}(x)) - v((1-\delta)x)\big) a^{ij} D_i\rho D_j\rho g(x)\, dS_x = 0 \tag{8.17}$$

for each $g \in L^2(B_2(x_1))$. Let $\mathcal{M} = \{g;\ \mathrm{supp}\ g \subset B_2(x_1), g = g_1 \text{ on } \partial Q \text{ for some } g_1 \in W^{1,2}(Q)\}$. Since $\mathcal{M}$ is dense in $L^2(B_2(x_1))$, it suffices to prove (8.17) for $g \in \mathcal{M}$. Taking

$$w(x) = \begin{cases} g_1(x)(\rho(x) - \delta) & \text{for } x \in Q_\delta, \\ 0 & \text{for } x \in Q - Q_\delta \end{cases}$$

as test function we get

(8.18)

$$\begin{aligned}
&\int_{B_2(x_1)} v(x_{\delta x_1}(x)) A^{ij} \frac{D_i\rho D_j\rho}{|D\rho|} g\, dS_x \\
&= \int_{Q_i} \Big[-D_i(A^{ij} g_1 D_j\rho)v + [a^{ij} - A^{ij}] D_i v g_1 D_j\rho + a^{ij} D_i v D_j g_1 (\rho - \delta) \\
&+ d^i D_i v g_1(\rho - \delta) + b^i v D_i g_1(\rho - \delta) + b^i v g_1 D_i\rho + c g_1(\rho - \delta) - g^i D_i(g_1(\rho - \delta)) \\
&- f_1 g_1(\rho - \delta)\Big]\, dx - \int_{B_2(x_1)} v(x_{\delta x_1}(x)) \Big[g_1(x_{\delta x_1}(x)\ A^{ij} \frac{D_i\rho D_j\rho}{|D\rho|}\Big|_{x = x_{\delta x_1}} \frac{dS_\delta}{dS} \\
&- g(x) A^{ij} \frac{D_i\rho D_j\rho}{|D\rho|}\Big]\, dS.
\end{aligned}$$

Let

$$\eta_{1,i} = \begin{cases} \rho(\frac{x}{1-\delta}) & \text{for } x \in Q_{i,\delta} \\ 0 & \text{for } x \in Q_i - Q_{i,\delta} \end{cases}$$

and let $\zeta_1 \in C^\infty(R_n)$ such that $\zeta_1(x) = 1$ for $x \in \mathcal{U}_1(x_1)$ and $\zeta_1(x) = 0$ for $x \in \bar{Q} - \mathcal{U}_2(x_1)$. Similarly, taking as a test function

$$w(x) = \begin{cases} \zeta_1(x) g_1(x) \eta_{1,i}(x) & \text{for } x \in Q_{i,\delta} \\ 0 & \text{for } x \in Q_i - Q_{i,\delta}, \end{cases}$$

we obtain

(8.19)

$$\begin{aligned}
&\int_{B_2(x_1)} v((1-\delta)x) A^{ij} \frac{D_i\rho D_j\rho}{|D\rho|} g(x)\, dS_x \\
&= (1-\delta)^{1-n} \int_{Q_i} \Big[-D_i\big(A^{ij} D_j\rho(\frac{x}{1-\delta})\big) g_1 + [a^{ij} - A^{ij}] D_i v g_1 D_j\rho(\frac{x}{1-\delta}) \\
&+ a^{ij} D_i v D_j \eta_{1,i} + d^i D_i v g_1 \eta_{1,i} + b^i v D_i g_1 \eta_{1,i} + b^i v g_1 D_i \eta_{1,i} + c g_1 \eta_{1,i} - f_1 g_1 \eta_{1,i} \\
&- g^i D_i(g_1 \eta_{1,i})\Big]\, dx \\
&+ \int_{B_2(x_1)} A^{ij} \frac{D_i\rho D_j\rho}{|D\rho|} \Big[A^{ij}(x) g(x) - A^{ij}((1-\delta)x) g_1((1-\delta)x)\Big] v((1-\delta)x)\, dS.
\end{aligned}$$

Letting $\delta \to 0$ and comparing the right sides of (8.18) and (8.19) we easily deduce (8.17). If we now modify the above argument, leading to (8.18) and (8.19), by replacing g_1 by v we arrive at

(8.20)
$$\lim_{\delta\to 0}\int_{B_2(x_1)} A^{ij}\frac{D_i\rho D_j\rho}{|D\rho|}v(x_{\delta x_1}(x))^2\,dS_x = \lim_{\delta\to 0}\int_{B_2(x_1)} A^{ij}\frac{D_i\rho D_j\rho}{|D\rho|}v((1-\delta)x)^2\,dS_x.$$

Finally, we prove that

$$\lim_{\delta\to 0}\frac{1}{2}\int_{B_2(x_1)} A^{ij}\frac{D_i\rho D_j\rho}{|D\rho|}v(x_{\delta x_1}(x))^2\,dS_x = \int_{B_2(x_1)} A^{ij}\frac{D_i\rho D_j\rho}{|D\rho|}(\zeta\varphi_i)^2\,dS_x. \tag{8.21}$$

To show this, we take as a test function in the equation $Lv = f_1 - D_i g^i$,

$$w(x) = \begin{cases} v(\alpha x)(\rho-\delta) & \text{for } x \in Q_\delta \cap \mathcal{U}_1(x_1) \\ 0 & \text{for } x \in Q_i - Q_\delta \cap \mathcal{U}_1(x_2), \end{cases}$$

where $\alpha \in [\alpha_1, 1]$, with α_1 chosen so close to 1 that $v(\alpha x) = 0$ on $Q_i - \mathcal{U}_1(x_1)$ for all $\alpha \in [\alpha_1, 1]$. Integrating by parts we obtain

$$\begin{aligned}
&\int_{Q_i\cap Q_\delta}\Big[a^{ij}D_ivD_jv(\alpha x) + b^ivD_iv(\alpha x) + d^iD_ivv(\alpha x)\\
&+ cv(\alpha x)\Big](\rho(x)-\delta)\,dx + \int_{Q_i\cap Q_\delta}[a^{ij}-A^{ij}]D_ivv(\alpha x)D_j\rho\,dx\\
&+ \int_{Q_i\cap Q_\delta} b^ivv(\alpha x)D_i\rho\,dx - \int_{Q_i\cap Q_\delta} D_i(A^{ij}D_j\rho)vv(\alpha x)\,dx\\
&- \int_{Q_i\cap Q_\delta} A^{ij}D_j\rho D_iv(\alpha x)v\,dx\\
&- \int_{\Gamma^1_{i,\delta}} A^{ij}\frac{D_i\rho D_j\rho}{|D\rho|}vv(\alpha x)\,dx = \int_{Q_i\cap Q_\delta} f_1v(\alpha x)(\rho(x)-\delta)\,dx\\
&+ \int_{Q_i\cap Q_\delta} g^iD_i(v(\alpha x)(\rho(x)-\delta))\,dx.
\end{aligned}$$

Letting $\delta \to 0$ we get

$$\begin{aligned}
&\int_{Q_i}\Big[a^{ij}D_ivD_jv(\alpha x) + b^ivD_iv(\alpha x) + d^iD_ivv(\alpha x)\\
&+ cv(\alpha x)\Big]\rho(x)\,dx + \int_{Q_i}[a^{ij}-A^{ij}]D_ivv(\alpha x)D_j\rho\,dx\\
&+ \int_{Q_i} b^ivv(\alpha x)D_i\rho\,dx - \int_{Q_i} D_i(A^{ij}D_j\rho)vv(\alpha x)\,dx - \int_{Q_i} A^{ij}D_j\rho D_iv(\alpha x)v\,dx\\
&- \int_{B_2(x_1)} A^{ij}\frac{D_i\rho D_j\rho}{|D\rho|}\varphi v(\alpha x)\,dx = \int_{Q_i} f_1v(\alpha x)\rho\,dx + \int_{Q_\delta} g^iD_i(v(\alpha x)\rho)\,dx.
\end{aligned}$$

We now observe that $v(\alpha x)$ is a solution in $W^{1,2}(Q_i)$ of the equation

$$D_i(a^{ij}(\alpha x)D_j v + b^i(\alpha x)v) + \alpha d^i(\alpha x)D_i v + \alpha^2 c(\alpha x)v = \alpha^2 f_1(\alpha x) + \alpha^2 D_i g^i(\alpha x)$$

and consequently

$$\int_{Q_i}\Big[a^{ij}(\alpha x)D_i v(\alpha x)D_j v + b^i(\alpha x)v(\alpha x)D_i v + \alpha d^i(\alpha x)D_i v(\alpha x)v$$
$$+ \alpha^2 c(\alpha x)v(\alpha x)v\Big]\rho(x)\,dx + \int_{Q_i} b^i(\alpha x)v(\alpha x)vD_i\rho\,dx$$
$$+ \int_{Q_i} a^{ij}(\alpha x)D_i v(\alpha x)vD_j\rho\,dx + \alpha^2\int_{Q_i} f_1(\alpha x)v\rho(x)\,dx + \int_{Q_i} g^i(\alpha x)D_i(v\rho)\,dx.$$

Adding the last two identities we obtain

(8.22)

$$\int_{Q_i}\Big[\big(a^{ij} + a^{ij}(\alpha x)\big)D_i v D_j v(\alpha x) + \big(b^i vD_i v(\alpha x) + b^i(\alpha x)v(\alpha x)D_i v\big)$$
$$d^i D_i vv(\alpha x) + \alpha d^i(\alpha x)D_i v(\alpha x)v + \big(c + \alpha^2 c(\alpha x)\big)vv(\alpha x)\Big]\rho(x)\,dx$$
$$+ \int_{Q_i} b^i vv(\alpha x)D_i\rho + b^i(\alpha x)v(\alpha x)vD_i\rho\,dx$$
$$+ \int_{Q_i}[a^{ij} - A^{ij}]D_i vv(\alpha x)D_j\rho\,dx - \int_{Q_i} D_i(A^{ij}D_j\rho)vv(\alpha x)\,dx$$
$$\int_{Q_i} A^{ij}D_j\rho D_i v(\alpha x)v\,dx + \int_{Q_i} a^{ij}(\alpha x)D_i v(\alpha x)vD_j\rho\,dx$$
$$- \int_{B_2(x_1)} A^{ij}\frac{D_i\rho D_j\rho}{|D\rho|}\varphi v(\alpha x)\,dx = \int_{Q_i}\big(f_1 v(\alpha x) + \alpha^2 f_1(\alpha x)v\big)\rho\,dx$$
$$+ \int_{Q_i}\big(g^i D_i(v(\alpha x)\rho) + \alpha^2 g^i(\alpha x)D_i(v\rho)\big)\,dx.$$

Since

$$\lim_{\alpha\to 1}\int_{Q_i}\Big[D_i\big(A^{ij}(\alpha x)D_j\rho\big) - D_i\big(A^{ij}(x)D_j\rho\big)\Big]vv(\alpha x)\,dx = 0$$

and

$$\lim_{\alpha\to 0}\int_{Q_i}\Big[A^{ij}(\alpha x) - A^{ij}(x)\Big]D_j\rho D_i v(\alpha x)v\,dx = 0,$$

we deduce from (8.22)

$$2\int_{Q_i}\Big[a^{ij}D_v D_j v + b^i vD_i v + d^i D_i vv + cv^2\Big]\rho(x)\,dx$$
$$+ 2\int_{Q_i}[a^{ij} - A^{ij}]D_i vvD_j\rho\,dx - \int_{B_2(x_1)} A^{ij}\frac{D_i\rho D_j\rho}{|D\rho|}\varphi\varphi_i\zeta\,dS_x$$
$$+ \int_{Q_i} d^i v^2 D_i\rho\,dx = 2\int_{Q_i} f_1 v\rho(x)\,dx + 2\int_{Q_i} g^i D_i(v\rho)\,dx.$$

Combining this with (8.20) and the fact that $\varphi = \zeta\varphi_i$ we obtain

$$\lim_{\delta\to 0}\int_{B_2(x_1)} A^{ij}\frac{D_i\rho D_j\rho}{|D\rho|}v(x_{\delta x_1}(x))^2\,dS_x = \int_{B_2(x_1)} A^{ij}\frac{D_i\rho D_j\rho}{|\rho|}(\varphi_i\zeta)^2\,dS_x.$$

Since $v(x_{\delta x_1}(x))$ converges weakly to $\varphi = \zeta\varphi_i$, the uniform convexity of L^2 implies that $v(x_{\delta x_1}(x)) \to \zeta\varphi_i$ in $L^2(\partial Q)$. We now observe that $\zeta \equiv 1$ on $B_2(x_1)$ and the relation (8.16) easily follows.

We close this chapter with the following remark. We have chosen in this chapter to work with domains having boundaries of class $C^{1,\alpha}$. Properties of a regularized distance, constructed in Theorem 8.2 suggest that this result should be true for boundaries of class $C^{1,\mathrm{Dini}}$. In fact, this only requires minor modifications of the arguments presented in this chapter. For more details we refer to a recent paper by Lieberman (see [**LI2**]). We shall return to this problem in Chapter 10, where we present a different approach to the $C^{1,\mathrm{Dini}}$–boundaries under more general assumption on the coefficients a^{ij}.

CHAPTER 9

THE SPACE $C_{n-1}(\bar{Q})$.

The results of Chapters 2 and 3 show that the weighted Sobolev space $\widetilde{W}^{1,2}(Q)$ arises in a natural way while solving the Dirichlet problem with L^2–boundary data. However, functions from this space may not have traces on the boundary of ∂Q and therefore the space $W^{1,2}(Q)$ is not suitable in our approach to the Dirichlet problem with L^2–boundary data. The purpose of this chapter is to construct a space $C_{n-1}(\bar{Q})$, which contains $W^{1,2}(Q)$, such that all elements of $C_{n-1}(\bar{Q})$ have traces in $L^2(\partial Q)$ on the boundary. Also, every function in $L^2(\partial Q)$ can be extended into $\bar{Q}$ to a function in $C_{n-1}(\bar{Q})$. The construction of the space $C_{n-1}(\bar{Q})$ presented in this chapter is due to Guščin [**GU**] and is obtained as the completion of the space $C(\bar{Q})$ with respect to a certain norm constructed with the aid of Carleson measures.

9.1. Spaces $C_{h,p}(\bar{Q})$ and Carleson measures.

Throughout this section we assume that Q is a bounded domain with ∂Q of class C^1.

Let $h : [0,\infty) \to [0,\infty)$ be an increasing, continuous function such that

$$\liminf_{r\to 0+} r^{-n}h(r) > 0. \tag{9.1}$$

For a bounded subset E of R_n we define

$$M_h(E) = \inf\{\sum_{i=1}^{\infty} h(r_i);\ E \subset \bigcup_{i=1}^{\infty} \overline{B(x^i,r_i)}\},$$

where the infimum is taken over all coverings $\{B(x^i,r_i)\}$ of the set E. The condition (9.1) implies that

$$|E|_e \leqq \text{Const}\, M_h(E) \quad \text{for all} \quad E \subset Q, \tag{9.2}$$

where $|\cdot|_e$ stands for the exterior Lebesgue measure. Let $v \in C(\bar{Q})$, we define the functional $\|v\|_{C_{h,p}}$ on $C(\bar{Q})$ by

$$\|v\|_{C_{h,p}} = \left[\int_{\bar{Q}} |v(x)|^p\, dM_h(x)\right]^{\frac{1}{p}} = \left[\int_0^{\infty} M_h(\{x \in \bar{Q};\ |v(x)|^p \geqq \lambda\})\, d\lambda\right]^{\frac{1}{p}},$$

where $1 \leqq p < \infty$. It is clear that $\|av\|_{C_{h,p}} = |a|\|v\|_{C_{h,p}}$ for each real number a. As an immediate consequence of (9.2) we obtain

$$\|v\|_p^p = \int_Q |v(x)|^p\, dx = \int_0^{\infty} |\{x \in Q;\ |v(x)|^p \geqq \lambda\}|\, d\lambda$$

$$\leqq \text{Const} \int_{\bar{Q}} |v(x)|^p\, dM_h(x) = \text{Const}\, \|v\|_{C_{h,p}}^p.$$

Consequently, if $\|v\|_{C_{h,p}} = 0$ then $v \equiv 0$ on Q. However, $\|\cdot\|_{C_{h,p}}$ does not satisfy the triangle inequality (see [GU]) . We proceed now to construct a norm on $C(\bar{Q})$ which is equivalent to the functional $\|\cdot\|_{C_{h,p}}$. This norm can be constructed using the concept of a Carleson measure. Due to this fact the completion of $C(\bar{Q})$ with respect to $\|\cdot\|_{C_{h,p}}$ has some nice properties. In particular, if $h(r) = r^{n-1}$, functions from this completion, denoted by $C_{n-1}(\bar{Q})$, have traces on the $(n-1)$–dimensional surfaces contained in $\bar{Q}$.

Let μ be a nonnegative Borel measure on R_n with support in Q. Following [CA] we impose the following condition on μ:

there exists a constant $C > 0$, such that for each ball $B(x,r)$

$$\mu(B(x,r)) \leqq Ch(r). \tag{9.3}$$

The measure μ satisfying (9.3) is called the Carleson measure. Setting

$$\|\mu\| = \inf\{C;\ C \text{ satisfies } (9.3)\}$$

we get $\mu(B(x,r)) \leqq \|\mu\| h(r)$ for each ball $B(x,r)$. For simplicity we always assume that $\mu(R_n) = \mu(\bar{Q}) = 1$. If $B(x,R_\circ) \supset Q$, then

$$1 = \mu(\bar{Q}) = \mu(B(x,R_\circ)) \leqq \|\mu\| h(R_\circ)$$

and consequently

$$\frac{1}{h(R_\circ)} \leqq \|\mu\|. \tag{9.4}$$

Let us denote the set of Carleson's measures μ, with $\mu(R_n) = \mu(\bar{Q}) = 1$, by $\mathcal{C}$.

For $v \in C(\bar{Q})$ we define

$$\|v\|_{\mathcal{C}} = \sup_{\mu \in \mathcal{C}} \left[\frac{1}{\|\mu\|} \int_{\bar{Q}} |v(x)|^p \, d\mu(x)\right]^{\frac{1}{p}}. \tag{9.5}$$

The functional $\|\cdot\|_{\mathcal{C}}$ defines a norm on $C(\bar{Q})$. Our first objective is to prove the equivalence of the functionals $\|\cdot\|_{\mathcal{C}}$ and $\|\cdot\|_{C_{h,p}}$.

We need the following lemma.

Lemma 9.1. *There exists a constant $c_\circ = c_\circ(n)$ such that for any decreasing sequence of bounded closed sets $\{E_j\}$, $j = 1, ..., q$, there exists a Carleson measure μ such that*

$$c_\circ M_h(E_s) \leqq \frac{1}{\|\mu\|} \mu(E_s) \tag{9.6}$$

for $s = 1, ..., q$.

PROOF. Let $K_\circ$ be a cube with its side equal to $l_\circ$ such that $K_\circ \supset E_1 \supset E_2 \supset ... \supset E_q$. We divide $K_\circ$ into 2^n equal cubes $K_1^1, ..., K_{2^n}^1$. The length of a side of each K_j^1 is equal to $l_1 = \frac{l_\circ}{2}$. Each of the cubes K_j^1 is divided into 2^n equal cubes and continuing this process, we obtain in the k-th step the partition $P_k = \{K_1^k, ..., K_{2^{kn}}^k\}$ of $K_\circ$, with the length of a side of each K_j^k equal to $l_k = \frac{l_{k-1}}{2} = \frac{l_\circ}{2^k}$. For a set $E \subset R_n$ we put

$$\Sigma_k(E) = \bigcup_{\{K \in P_k,\, K \cap E \neq \varnothing\}} K$$

and $P = \bigcup_{k=1}^{\infty} P_k$.
For a fixed k, we define a measure μ_k°, with supp $\mu_k^\circ \subset \Sigma_k(E_1)$, such that

$$\mu_k^\circ(G) = \frac{h(l_k)}{(l_k)^n} |G \cap \Sigma_k(E_1)|$$

for all measurable sets $G \subset R_n$. This means that the density of μ_k° is equal to $\frac{h(l_k)}{(l_k)^n}$ on each $K_j^k \in \Sigma_k(E_1)$. Starting with measure μ_k° we construct a measure μ_k having the following properties
(i) for each $K_i \in \bigcup_{m=0}^{k} P_m$, where $P_\circ = \{K_\circ\}$, $\mu_k(K_i) \leqq h(l_i)$,
(ii) for every E_s, $s = 1, ..., q$, there exists a covering $\{K_i\} \subset \bigcup_{m=0}^{k} P_m$ of the set E_s, such that $\mu_k(\Sigma_k(E_s)) = \sum_i \mu_k(K_i) = \sum_i h(l_i)$.
To obtain μ_k we construct a finite sequence of measures $\{\mu_k^i\}$, $i = 1, ..., k$, the measure μ_k^i being obtained by modifying the measure μ_k^{i-1}, $i = 1, ..., k$. Then the measure μ_k having the properties (i) and (ii) is obtained by setting $\mu_k = \mu_k^k$.
We choose a cube $K_{j_1} = K_{j_1}^{k-1} \in P_{k-1}$. If $\mu_k^\circ \leqq h(l_{k-1})$ we define $\mu_k^1(K_{j_1}) = \mu_k^\circ(K_{j_1})$. If $\mu_k^\circ(K_{j_1}) > h(l_{k-1})$, we consider cubes K_i in K_{j_1} belonging to $\Sigma_k(E_1) - -\Sigma_k(E_2)$. We now multiply the density of μ_k° on these cubes by the same positive number less than 1, to get the measure of K_{j_1} equal to $h(l_{k-1})$. If this is not possible, we decrease the density on these cubes to 0. If the equality is not achieved, we change the density on cubes $K_i \subset K_{j_1}$ belonging to $\Sigma_k(E_2) - \Sigma_k(E_3)$. We continue this process until we convert the measure of K_{j_1} to $h(l_{k-1})$. We observe that if $K_{i_\circ} \subset K_{j_1}^{k-1} \cap \Sigma_k(E_s)$ for some $1 < s \leqq q$, then the modified density is less than the original density of μ_k°, and this density is 0 on all cubes $K_i^k \subset K_{j_1}^{k-1}$ not belonging to $\Sigma_k(E_s)$; so the measure of K_{j-1} is equal to $h(l_{k-1})$. Applying this procedure to all cubes $K_{j_1}^{k-1}$, $j_1 = 1, ..., 2^{n(k-1)}$, we obtain a new measure μ_k^1. This measure has the following properties:
For all cubes K_i^k, $i = 1, ..., 2^{kn}$, and K_j^{k-1}, $j = 1, ..., 2^{n(k-1)}$, we have

$$\mu_k^1(K_i^k) \leqq h(l_k) \quad \text{and} \quad \mu_k^1(K_j^{k-1}) \leqq h(l_{k-1})$$

and moreover for each $s = 1, ..., q$ and each cube $K_{i_\circ}^k \subset \Sigma_k(E_s)$,

$$\text{either} \quad \mu_k^1(K_{i_\circ}^k) = h(l_k) \quad \text{or} \quad \mu_k^1(K_{j_1} \cap \Sigma_k(E_s)) = h(l_{k-1}),$$

where $K_{j_1} = K_{j_1}^{k-1} \supset K_{i_\circ}^k$.
We now construct a measure μ_k^2 in a similar way. We select a cube $K_{j_2} = K_{j_2}^{k-2} \in P_{k-2}$. If $\mu_k^1(K_{j_2}) \leqq h(l_{k-2})$, we define $\mu_k^2(K_{j_2}) = \mu_k^1(K_{j_2})$. If $\mu_k^1(K_{j_2}) > h(l_{k-2})$, we first multiply the density of the measures μ_k^1 on cubes $K_i^k \subset \overline{K_{j_2} - \Sigma_k(E_2)}$ by the same

constant less than 1, to get the measure of K_{j_2} equal to $h(l_{k-2})$. If this is not possible, we decrease the density on these cubes to 0. If, as a result, the equality is not achieved, we also decrease the density on cubes $K_i^k \subset \overline{K_{j_2} - \Sigma(E_3)}$ and so on. Applying this procedure to all cubes $K_{j_2} \subset P_{k-2}$, we obtain a measure μ_k^2. The measure μ_k^2 has the following properties:

$$\mu_k^2(K_i) \leqq h(l_k) \quad \text{for all} \quad K_i \in P_k \cup P_{k-1} \cup P_{k-2},$$

for each $s = 1, ..., q$ and every cube $K_{i_o} = K_{i_o}^k \subset \Sigma_k(E_s)$, either

$$\mu_k^2(K_{i_o}^k) = h(l_k),$$

or

$$\mu_k^2(K_{j_1}^{k-1} \cap \Sigma_k(E_s)) = h(l_{k-1}),$$

where $K_{j_1}^{k-1} \supset K_{i_o}$, or

$$\mu_k^2(K_{j_2}^{k-2} \cap \Sigma_k(E_s)) = h(l_{k-2}),$$

where $K_{j_2}^{k-2} \supset K_{j_o}$.
Indeed, if $\mu_k^2(K_{i_o}) = \mu_k^1(K_{i_o})$, then according to our construction of μ_k^2, we have $\mu_k^2(K_i^k) = \mu_k^1(K_i^k)$ for all $K_i^k \subset K_{j_2}^{k-2} \cap \Sigma_k(E_s)$. Then, either $\mu_k^1(K_{i_o}) = h(l_k)$, that is, $\mu_k^2(K_{i_o}) = h(l_k)$ or $\mu_k^1(K_{j_1}^{k-1} \cap \Sigma_k(E_s)) = h(l_{k-1})$, that is,

$$\mu_k^2(K_{j_1}^{k-1} \cap \Sigma_k(E_s)) = \sum_{K_i^k \subset K_{j_1}^{k-1} \cap \Sigma_k(E_s)} \mu_k^2(K_i^k) = \mu_k^1(K_{j_1}^{k-1} \cap \Sigma_k(E_s)) = h(l_{k-1}).$$

If $\mu_k^2(K_{i_o}) < \mu_k^1(K_{i_o})$, then $\mu_k^2(K_{j_2}^{k-2})$ and for all $K_i^k \subset \overline{K_{j_2}^{k-2} - \Sigma_k(E_s)}$ we have $\mu_k^2(K_i^k) = 0$. Consequently,

$$\mu_k^2(K_{j_2}^{k-2}) = \mu_k^2(K_{j_2}^{k-2} \cap \Sigma_k(E_s)) = h(l_{k-2}).$$

Similarly we construct measures $\mu_k^3, ..., \mu_k^k$ and set $\mu_k = \mu_k^k$. This measure satisfies (i). As in the case of the second property of the measure μ_k^2, we show that for each $s = 1, ..., q$ and every $K_o = K_{i_o}^k \subset \Sigma_k(E_s)$, there exists an integer m, with $0 \leqq m \leqq k$, such that for a cube $K_{j_m}^{k-m} \supset K_{i_o}$ (if $m = 0$, $K_{j_o}^2 = K_{i_o}$) we have

$$\mu_k(K_{j_m}^{k-m} \cap \Sigma_k(E_s)) = \mu_k(K_{j_m}^{k-m}) = h(l_{k-m}), \tag{9.7}$$

Since the construction leading to (9.7) can be carried out for every cube $K_{i_o}^k$ in $\Sigma_k(E_s)$, $s = 1, ..., q$, we obtain a covering of E_s satisfying (ii). Obviously, we only choose cubes which do not intersect at interior points.
We now select from the sequence $\{\mu_k\}$, a weakly convergent subsequence $\{\mu_{k_m}\}$ to a measure μ. We relabel this subsequence again by μ_k. We now show that μ satisfies (9.4). It follows from the property (i) that for each ball $B(x,r)$ we have

$$\mu(B(x,r)) \leqq \limsup_{k\to\infty} \mu_k(\Sigma_{k_o}(B(x,r)) \leqq C_1 h(l_{k_o}) \leqq C_1 h(r),$$

where $k_\circ$ is such that $l_{k_\circ} \leqq r < 2l_{k_\circ}$ and $C_1 > 0$ is a constant depending only on n. Consequently $\|\mu\| \leqq C_1$. On the other hand, by virtue of the property (ii), $\mu_k(\Sigma_k(E_s)) \geqq c_1 M_h(E_s)$ for $k = 1, 2, ...$ and $s = 1, ..., q$, where $c_1 > 0$ is a constant depending on n. Consequently for every open set $G_s \supset E_s$, we have

$$\mu(G_s) \geqq \limsup_{k\to\infty} \mu_k(\Sigma_k(E_s)) \geqq c_1 M_h(E_s)$$

and

$$\mu(E_s) = \inf_{G_s \supset E_s} \mu(G_s) \geqq c_1 M_h(E_s).$$

The inequality $\|\mu\| \leqq c_1$ implies the estimate (9.6).

In the case where $q = 1$, this result is due to Frostman [**FR**] (see also [**CA1**]).

Theorem 9.1. *There exists a positive constant $c = c(n,p)$ such that*

$$c\|v\|_{C_{h,p}} \leqq \|v\|_{\mathcal{C}} \leqq \|v\|_{C_{h,p}} \tag{9.8}$$

for all $v \in C(\bar{Q})$.

PROOF. The proof of the right hand side of (9.8) is straightforward. Let $\mu \in \mathcal{C}$, then

$$\|v\|^p_{p,\mu} = \int_Q |v(x)|^p \, d\mu(x) = \int_0^\infty \mu(\{x \in \bar{Q};\, |v(x)| > \lambda\})\, d\lambda^p.$$

It follows from the definition of the Carleson measure that $\mu(E) \leqq \|\mu\| M_h(E)$. Hence

$$\|v\|^p_{p,\mu} \leqq \|\mu\| \int_0^\infty M_h(\{x \in \bar{Q};\, |v(x)|^p \geqq\})\, d\lambda^p$$
$$= \|\mu\| \int_Q |v(x)|^p \, dM_h = \|\mu\| \|v\|^p_{C_{h,p}}$$

and this yields the estimate

$$\sup_{\mu \in \mathcal{C}} \frac{1}{\|\mu\|^{\frac{1}{p}}} \|v\|_{p,\mu} \leqq \|v\|_{C_{h,p}}.$$

To prove the left hand side of (9.8), we use Lemma 9.1. Let $0 = \lambda_\circ < \lambda_1 < ... < \lambda_q = \|v\|^p_{C(\bar{Q})}$ be a partition of the interval $[0, \|v\|^p_{C(\bar{Q})}]$. We now define a finite sequence $\{E_s\}$ of closed bounded sets by $E_s = \{x \in \bar{Q};\, |v(x)| \geqq \lambda_s\}$, $s = 1, ..., q$. By Lemma 9.1, there is a Carleson measure μ such that

$$\sum_{s=1}^{q} M_h(E_s)(\lambda_s - \lambda_{s-1}) \leqq \frac{1}{c_\circ \|\mu\|} \sum_{s=1}^{q} \mu(E_s)(\lambda_s - \lambda_{s-1})$$
$$\frac{1}{c_\circ \|\mu\|} \int_0^a \mu(\{x \in \bar{Q};\, |v(x)|^p \geqq \lambda\})\, d\lambda = \frac{1}{c_\circ \|\mu\|} \|v\|^p_{p,\mu},$$

where $a = \|v\|^p_{C(\bar{Q})}$. This completes the proof.

9.2. Spaces $\widetilde{C}_{h,p}$ and C_{n-1}.

Let us now denote by $\widetilde{C}_{h,p}(\bar{Q})$ the completion of $C(\bar{Q})$ with respect to $\|\cdot\|_{h,p}$. If $h(r) = r^{n-1}$ and $p = 2$ we use notation $C_{n-1}(\bar{Q})$. The corresponding norms in $\widetilde{C}_{h,p}(\bar{Q})$ and $C_{n-1}(\bar{Q})$ are denoted by $\|\cdot\|_{\widetilde{C}_{h,p}}$ and $\|\cdot\|_{C_{n-1}}$, respectively. The fact that $\|\cdot\|_C$ and $\|\cdot\|_{C_{h,p}}$ are equivalent leads to the existence of traces of functions from the space $\widetilde{C}_{h,p}(\bar{Q})$: namely, let $v \in \widetilde{C}_{h,p}(\bar{Q})$ and let $\mu \in C$; then there exists a sequence $\{v_m\}$ in $C(\bar{Q})$ such that $\|v_m - v\|_{C_{h,p}} \to 0$ as $m \to \infty$. By Theorem 9.1, this sequence converges to a function w in $L^p(Q, \mu)$. We call w the trace of v on the support of μ. From now on, the trace function w corresponding to v will be denoted by the same letter v. Since $\|v\|^p_p \leqq \text{Const}\,\|v\|_{C_{h,p}}$ for each $v \in C(\bar{Q})$, the space $\widetilde{C}_{p,h}(\bar{Q})$ is continuously imbedded in $L^p(Q)$.

Let μ be a measure with support on the intersection of a $(n-1)$–dimension al hyperplane H with $\bar{Q}$. If this measure coincides with the $(n-1)$–dimensional Lebesgue measure, then μ is the Carleson measure corresponding to $h(r) = r^{n-1}$ and its norm is independent of the choice of h. This implies that the traces of v on $H \cap \bar{Q}$ are continuous with respect to translation of H; that is

$$C_{n-1}(\bar{Q}) \subset C([a,b], L^2(\Omega)),$$

where $[a,b] \times \Omega \subset Q$. In the next section we shall return to the problem of traces in $C_{h,p}$.

It is also clear that we can define traces on the $(n-1)$–dimensional smooth surfaces in Q. In particular, every v in $C_{n-1}(\bar{Q})$ has a trace ϕ on ∂Q, which belongs to $L^2(\partial Q)$. The function ϕ can be recovered in the following way:

To every $x^\circ \in \partial Q$, corresponds a neighbourhood V_{x° in ∂Q of the point x°, such that

$$\lim_{\delta \to 0} \int_{V_{x^\circ}} [u(x + \delta\nu(x^\circ)) - \phi(x)]^2 \, dS_x = 0, \tag{9.9}$$

where $\nu(x^\circ)$ is an inward normal to ∂Q at x°. This relation will be used to recover a boundary data in our approach to the Dirichlet problem developed in the next chapter.

We point out here that the continuity of v on Q, together with the relation (9.9), do not imply that $v \in C_{n-1}(\bar{Q})$. As an example consider

$$v(x_1, x_2) = \begin{cases} \frac{x_1^2 - |x_2|}{x_1^{\frac{5}{2}}}(1 - x_1) & \text{for } 0 < x_1 < 1,\ |x_2| < x_1^2, \\ 0 & \text{elsewhere in } R_2. \end{cases}$$

It is clear that $v \in C(B((1,0),1))$, $\lim_{r\to 1} \int_0^{2\pi} rv(1 + r\cos\phi, r\sin\phi)^2 \, d\phi$ exists and the integral $\int_0^1 v(x_1, 0)^2 \, dx_1$ is divergent.

9.3. $W^{1,2}(Q) \subset C_{n-1}(\bar{Q})$.

We begin with a property of the space $C_{n-1}(\bar{Q})$, which shows its relation to the space $W^{1,2}(Q)$ and consequently its importance in our approach to the Dirichlet problem with L^2–boundary data.

Theorem 9.2. *The space $W^{1,2}(Q)$ is continuously embedded in $C_{n-1}(\bar{Q})$ and the set of traces on ∂Q of functions in $C_{n-1}(\bar{Q})$ coincides with $L^2(\partial Q)$.*

PROOF. To prove the first part we take $v \in C^\infty(\bar{Q})$ and set $E_\lambda = \{x \in Q;\ v(x)^2 > \lambda\}$. It follows from Theorem 2 on p. 33 in [MA] that there exists a covering of $\{E_\lambda\}$ with balls $\{B(x^i, r_i)\}$ such that

$$\sum_{i=1}^{\infty} r_i^{n-1} \leqq C(n)|\partial E_\lambda|,$$

where $C(n) > 0$ is a constant depending only n. Consequently $M_{n-1}(E_\lambda) \leqq C(n)|\partial E_\lambda|$ and by the formula on p. 37 [MA] we have

$$\begin{aligned} \int_{\bar{Q}} v(x)^2\, dM_{n-1}(x) &= \int_0^\infty M_{n-1}(\bar{E}_\lambda)\, d\lambda \leqq C_1(n) \int_0^\infty |\partial E_\lambda|\, d\lambda \\ &= 2C_1(n) \int_Q |Dv(x)||v(x)|\, dx \leqq C_1(n)\|v\|_{1,2}^2 \end{aligned}$$

and the use of the density of $C^\infty(\bar{Q})$ in $W^{1,2}(Q)$ completes the proof of the embedding $W^{1,2}(Q) \subset C_{n-1}(\bar{Q})$.

Concerning the second part of our assertion we prove a stronger result. Let $\Gamma = \bar{\Gamma}$ be a smooth $(n-1)$–dimensional surface in $\bar{Q}$. By using a partition of unity and flattening Γ locally, we may reduce our situation to the case where $\Gamma \subset \{x_n;\ x_n = 0\}$. Let $\phi \in L^2(\Gamma)$. We extend ϕ by 0 outside Γ. Let $\{w_k\}$ be a sequence of continuous functions with supports in $\{x' \in R_{n-1},\ (x', 0) \in \bar{Q}\}$ and converging to ϕ in $L^2(R_{n-1})$. We define functions $v_k \in C(R_n)$, $k = 1, 2, \ldots$, by

$$v_k(x', x_n) = \begin{cases} w_k(x') & \text{for } x_n = 0, \\ \frac{1}{|B(x', |x_n|)|} \int_{B(x', |x_n|)} w_k(\xi')\, d\xi' & \text{for } x_n \neq 0. \end{cases}$$

For an arbitrary and bounded set $E \subset R_{n-1}$, we get

$$W_E = R_n - \bigcup_{\xi' \notin E} \{(x', x_n);\ |x' - \xi'| < |x_n|\}.$$

It is clear that $(x', x_n) \in W_E$ if and only if $B(x', |x_n|) \subset E$. It is easy to show that

$$M_{n-1}(W_E) \leqq C(n)|E|. \tag{9.10}$$

We now observe that the inequality $|v_k(x', x_n) - v_m(x', x_n)| > t$ implies that $M(w_k - w_m(\xi) > t$ for $\xi \in B(x', |x_n|)$, where $M(w_k - w_m)$ denotes the Hardy-Littlewood maximal function of $|w_k - w_m|$ (see Chap.4, Section 4.3). Consequently

$$\{x \in \bar{Q};\ |v_k(x) - v_m(x)|^2 > \lambda\} \subset W_{A(\lambda)},$$

where $A(\lambda) = \{x' \in R_{n-1};\ M(w_k - w_m)(x')^2 > \lambda\}$ and using (9.10) we obtain the following estimate

$$\begin{aligned} \|v_k - v_m\|_{C_{n-1}(\bar{Q})}^2 &= \int_0^\infty M_{n-1}(\{x \in \bar{Q};\ |v_k(x) - v_m(x)|^2 > \lambda\})\, d\lambda \\ &\leqq \int_0^\infty M_{n-1}(W_{A(\lambda)})\, d\lambda \leqq C(n) \int_0^\infty |A(\lambda)|\, d\lambda \\ &= C(n)\|M(w_k - w_m)\|_2^2. \end{aligned}$$

The application of the classical inequality

$$\|M(w_k - w_m)\|_2 \leqq \text{Const}\, \|w_k - w_m\|_2$$

completes the proof.

9.4. Properties of traces of functions in $\tilde{C}_{h,p}$.

A trace of a function in $\widetilde{C}_{h,p}(\bar{Q})$ can be defined on an arbitrary closed subset $\Gamma \subset \bar{Q}$ with positive Hausdorff measure. Indeed, let

$$|\Gamma|_h = \lim_{\delta \to 0} \inf\Big(\{\sum_i h(r_i);\ \Gamma \subset \bigcup_i B(x^i, r_i), r_i < \delta\}\Big),$$

where inf is taken over all coverings of Γ with balls $B(x^i, r_i)$. Here $|\cdot|_h$ stands for the Hausdorff measure. Obviously we have $|\Gamma|_h = M_h(\Gamma)$. Suppose that $|\Gamma|_k > 0$ and let $v \in \widetilde{C}_{h,p}(\bar{Q})$. For a sequence $\{v_k\}$ in $C(\bar{Q})$ converging in $C_{h,p}(\bar{Q})$ to v, we have

$$\begin{aligned}\|v_k - v_m\|_{C_{h,p}(\Gamma)} &= \int_0^\infty M_h(\{x \in \Gamma;\ |v_k(x) - v_m(x)|^p > \lambda\})\, d\lambda \\ &\leqq \|v_k - v_m\|^p_{C_{h,p}}.\end{aligned}$$

Consequently $\{v_k\}$ restricted to Γ, has a limit in $\widetilde{C}_{h,p}(\Gamma)$. This limit, denoted again by v, is a trace of v on Γ; in particular, if $|\Gamma| > 0$, where $|\cdot|$ denotes the $(n-1)$–dimensional Lebesgue measure corresponding to $h(r) = r^{n-1}$. Consequently, if $v \in C_{n-1}(\Gamma)$, then its trace belongs to $L^2(\Gamma)$. The converse to the last remark is contained in the second part of Theorem 9.2.

We conclude this section by proving some properties of traces of functions from $\widetilde{C}_{h,p}(\bar{Q})$. These properties show that traces change continuously under smooth deformations of surfaces. In particular, we derive a necessary and sufficient condition for v to belong to $\tilde{C}_{h,p}$.

Theorem 9.3.. *Let $v \in \widetilde{C}_{h,p}(\bar{Q})$ and let μ be a Borel measure on R_{2n} with $\operatorname{supp}\mu \subset \bar{Q} \times \bar{Q}$ and $\mu(R_{2n}) = 1$. Let*

$$\mu_1(G) = \mu(G \times R_n) \text{ and } \mu_2(G) = \mu(R_n \times G)$$

for all measurable sets $G \subset R_n$. Suppose that μ_1 and μ_2 are Carleson's measures. Then for every $\epsilon >$ there exist $\delta > 0$ and $\sigma > 0$ such that

$$\frac{1}{\max(\|\mu_1\|, \|\mu_2\|)} \iint_{R_{2n}} |v(x) - v(y)|^p\, d\mu(x,y) < \epsilon, \tag{9.11}$$

whenever

$$\mu(\{(x,y) \in R_{2n};\ |x - y| \geqq \delta\}) < \sigma. \tag{9.12}$$

PROOF. Let $\{v_k\}$ be a sequence in $C(\bar{Q})$ converging to v in $C_{h,p}(\bar{Q})$. We fix k so that

$$\sup_{\mu \in C} \frac{1}{\|\mu\|} \|v - v_k\|_{p,\mu}^p < \frac{\epsilon}{4^p}$$

and let $A = \max(\|\mu_1\|, \|\mu_2\|)$. The numbers $\sigma > 0$ and $\delta > 0$ are determined by conditions

$$\|v_k\|_{C(\bar{Q})}^p \sigma < \frac{A\epsilon}{8^p}$$

and

$$|v_k(x) - v_k(y)| < \frac{\epsilon^{\frac{1}{p}} A^{\frac{1}{p}}}{4}$$

whenever $|x - y| < \delta$. Thus if μ satisfies (9.12) we have the following estimate

$$\begin{aligned}
&\iint_{R_{2n}} |v(x) - v(y)|^p \, d\mu(x,y) \leqq \iint_{R_{2n}} [|v(x) - v_k(x)| \\
&+ |v_k(x) - v_k(y)|]^2 \, d\mu(x,y) \leqq \left[\left(\int_{R_n} |v(x) - v_k(x)|^p \, d\mu_1(x) \right)^{\frac{1}{p}} \right. \\
&\left. + \left(\iint_{R_{2n}} |v_k(x) - v_k(y)|^p \, d\mu(x,y) \right)^{\frac{1}{p}} + \left(\int_{R_n} |v_k(y) - v(y)|^p \, d\mu_2(y) \right)^{\frac{1}{p}} \right]^p \\
&\leqq \left[\frac{\epsilon^{\frac{1}{p}} \|\mu_1\|^{\frac{1}{p}}}{4} + 2 \left(\iint_{(x,y) \in R_{2n};\, |x-y| \geqq \delta} \|v_k\|_{C(\bar{Q})}^p \, d\mu(x,y) \right)^{\frac{1}{p}} \right. \\
&\left. + \left(\iint_{(x,y) \in R_{2n};\, |x-y| < \delta} \frac{\epsilon}{4^p} \, d\mu(x,y) \right)^{\frac{1}{p}} + \frac{\epsilon^{\frac{1}{p}} \|\mu_2\|^{\frac{1}{p}}}{4} \right]^p \\
&\leqq \left[\max(\|\mu_1\|^{\frac{1}{p}}, \|\mu_2\|^{\frac{1}{p}}) \frac{\epsilon^{\frac{1}{p}}}{2} + 2\|v_k\|_{C(\bar{Q})} \sigma^{\frac{1}{p}} + \frac{\epsilon^{\frac{1}{p}} A^{\frac{1}{p}}}{4} \right]^p \leqq A\epsilon
\end{aligned}$$

and the result follows.

Theorem 9.3 can be interpreted in the following way: traces of a function $v \in \widetilde{C}_{h,p}(\bar{Q})$ on supports of Carleson's measures μ_1 and μ_2 are close in the sense of inequality (9.11), provided the measures μ_1 and μ_2 are sufficiently close. The closeness of μ_1 and μ_2 is expressed by the condition (9.12).

We now apply this theorem to a measure with support on a graph of a mapping of $\bar{Q}$ into $\bar{Q}$.

Let ν be a Carleson measure with $\operatorname{supp} \nu \subset \bar{Q}$, $\nu(\bar{Q}) = \nu(R_n) = 1$ and let $\psi : R_n \to R$ be measurable mapping such that $\psi(\operatorname{supp} \nu) \subset Q$. We define a measure ν_ψ by

$$\nu_\psi(G) = \nu(\psi^{-1}(G))$$

for all measurable sets $G \subset R_n$.

Theorem 9.4. *Let $v \in \widetilde{C}_{h,p}(\bar{Q})$. Then for every $\epsilon > 0$ and $\lambda > 0$ there exist $\delta > 0$ and $\sigma > 0$ such that for every Carleson measure ν and every measurable mapping ψ : supp $\nu \to \bar{Q}$ (supp $\nu \subset \bar{Q}$) we have*

$$\frac{1}{\|\nu\|}\int_{R_n} |v(\psi(x)) - v(x)|^p \, d\nu(x) < \epsilon, \tag{9.13}$$

whenever

$$\nu(\{x \in R_n;\ |\psi(x) - x| \geqq \delta\}) < \sigma \tag{9.14}$$

and

$$\|\nu_\psi\| \leqq \lambda \|\nu\|. \tag{9.15}$$

PROOF. We define a measure μ on R_{2n} by

$$\mu(E) = \nu(\{x \in R_{2n};\ (x, \psi(x)) \in E\})$$

for all measurable sets $E \subset R_{2n}$. Then $\mu(R_{2n}) = \nu(R_n) = 1$ and supp $\mu \subset \bar{Q} \times \bar{Q}$ and for measurable sets $G \subset R_n$ we have

$$\mu_1(G) = \mu(G \times R_n) = \nu(\{x \in R_n;\ (x, \psi(x)) \in G \times R_n\}) = \nu(G),$$
$$\mu_2(G) = \mu(R_n \times G) = \nu(\{x \in R_n;\ (x, \psi(x)) \in R_n \times G\}) = \nu(\psi^{-1}(G)) = \nu_\psi(G)$$

and by (9.14)

$$\mu(\{(x,y) : |x - y| \geqq \delta\}) = \nu(\{x \in R_n;\ |x - \psi(x)| \geqq \delta\}) < \sigma,$$

that is (9.12) holds. Applying Theorem 9.3 we get

$$\frac{1}{\|\nu\|}\int_Q |v(x) - v(x, \psi(x))|^p \, d\nu \leqq \frac{\max(1, \lambda)}{\max(\|\nu\|, \|\nu_\psi\|)} \int_{R_{2n}} |v(x) - v(y)|^p \, d\mu(x,y) < \epsilon.$$

Here the choice of $\delta > 0$ and $\sigma > 0$ depends on $\frac{\epsilon}{\max(1,\lambda)}$.

If we restrict ourselves to the mappings ψ satisfying (9.14) and (9.15) for all measures ν with $\nu(R_n) = 1$ and with supp $\psi \subset \bar{Q}$, then the estimate (9.13) holds for $v \circ \psi - v$, that is, we have the

COROLLARY. *Let $v \in C_{n-1}(\bar{Q})$. For all numbers $\epsilon > 0$ and $\lambda > 0$ there exists $\delta = \delta(\epsilon, \lambda)$ such that for every C^1–diffeomorphism $\psi : \bar{Q} \to \bar{Q}$ such that $\|\psi\|_{C^1(\bar{Q})} \leqq \lambda$ and $\sup_{x \in \bar{Q}} |\psi(x) - x| < \delta$ we have $\|v - v \circ \psi\|_{C_{n-1}} < \epsilon$.*

To show the second application of Theorem 9.3, we associate with a given Carleson measure μ a measure μ_ρ given by

$$\mu_\rho(E) = \int_Q \frac{|E(x) \cap Q_\rho(x)|}{|Q_\rho(x)|} \, d\mu(x)$$

for all measurable subsets $E \subset R_{2n}$, where $Q_\rho(x) = Q \cap B(x, \rho)$ and $E(x) = \{y \in R_n;\ (x,y) \in E\}$.

Theorem 9.5. *Suppose that the function $h : [0,\infty) \to [0,\infty)$ (see (9.1)) satisfies the condition:*

There exists a constant $d_\circ > 0$ such that

$$r_2^{-n} h(r_2) \leqq d_\circ r_1^{-n} h(r_1)$$

for all $0 < r_1 < r_2$.

Let $v \in \widetilde{C}_{h,p}(\bar{Q})$. Then for every $\epsilon > 0$ there exists a $\delta > 0$ such that

$$\|v - v_\rho\|_{C_{h,p}(\bar{Q})} < \epsilon,$$

whenever $\rho \in (0,\delta)$ and $v_\rho(x) = \frac{1}{|Q_\rho(x)|}\int_{Q_\rho(x)} v(y)\,dy$.

PROOF. First we observe that by the regularity assumption on ∂Q, there exists $c_\circ > 0$ such that for all $x \in \bar{Q}$

$$|Q_\rho(x)| \geqq c_\circ \rho^n. \tag{9.17}$$

As in Theorem 9.3 we associate with μ_ρ, projection measures $\mu_1 = (\mu_\rho)_1$ and $\mu_2 = (\mu_\rho)_2$. We have $\mu_1 = \mu$. We commence by showing that μ_2 is a Carleson measure. Indeed, let B_r be a ball of radius r in R_n and consider

$$\mu_2(B_r) = \mu_\rho(R_n \times B_r) = \int_Q \frac{|B_r \cap Q_\rho(x)|}{|Q_\rho(x)|}\,d\mu(x).$$

Since the integration is taken over the intersection of $\bar{Q}$ with a ball of radius $r+\rho$, the relations (9.16) and (9.17) imply that

$$\mu_2(B_r) \leqq C(Q)\frac{\min(r^n,\rho^n)}{\rho^n}\|\mu\| h(r+\rho) \leqq d_\circ 2^n C(Q)\|\mu\| h(r)$$

if $\rho \leqq r$, and

$$\mu_2(B_r) \leqq C(Q)\frac{r^n}{\rho^n}\|\mu\| d_\circ 2^n h(\rho) \leqq 2^n d_\circ^2 C(Q) h(r)$$

if $r < \rho$. Hence $\|\mu\|_2 \leqq \text{Const}\,\|\mu\|$. Since for $\rho \in (0,\delta)$

$$\mu_\rho(\{(x,y) : |x-y| \geq \delta\}) = 0,$$

according to Theorem (9.3) we have

$$\begin{aligned}
&\frac{1}{\|\mu\|}\int_{\bar{Q}} |v(x) - \frac{1}{|Q_\rho(x)|}\int_{Q_\rho} v(y)\,dy|^p\,d\mu(x)\\
&\leqq \frac{\text{Const}}{\max(\|\mu\|,\|\mu\|)}\int_Q \frac{1}{|Q_\rho(x)|}\int_{Q_\rho(x)} |v(x)-v(y)|^p\,dy\,d\mu(x)\\
&= \frac{\text{Const}}{\max(\|\mu\|,\|\mu_2\|)}\int_{R_{2n}} |v(x)-v(y)|^p\,d\mu_\rho(x,y) < \text{ Const } \epsilon.
\end{aligned}$$

Theorem 9.5 establishes a relation between the trace of a function $v \in \widetilde{C}_{h,p}(\bar{Q})$ on a support of a given Carleson measure and a function $v_\rho \in C(\bar{Q})$ which approximates v in $C_{h,p}(\bar{Q})$. As we previously observed, $\widetilde{C}_{h,p}(\bar{Q})$ is continuously embedded in $L^p(Q)$. Now, due to the fact that $v_\rho \in C(\bar{Q})$ for $v \in L^p(Q)$, Theorem 9.5 gives necessary and sufficient condition for $v \in L^p(Q)$ to belong to $\widetilde{C}_{h,p}(Q)$, namely:

Theorem 9.6. *A function $v \in L^p(Q)$ belongs to $\widetilde{C}_{h,p}(\bar{Q})$ if and only if v_ρ converges to v in $C_{h,p}(\bar{Q})$.*

CHAPTER 10
C_{n-1}–ESTIMATE OF THE SOLUTION OF THE DIRICHLET PROBLEM WITH L^2–BOUNDARY DATA

The interesting feature of the space C_{n-1}, constructed in Chapter 9, is that the C_{n-1}–norm of the solution of the Dirichlet problem can be estimated by the L^2–norms of the boundary data and right hand side of the equation. To derive this estimate we use some properties of the non tangential maximal function and the area integral that can be associated with a solution of the equation. These two quantities have been known for harmonic functions (see [**ST**] and [**BG**]). In this chapter we closely follow the recent results obtained by Guščin ([**GU**] see also [**ŠE**]). Another aspect of the Guščin approach is the significant weakening of the regularity of the leading coeficients a^{ij}. We only require the Dini condition on the boundary ∂Q. His method of proving the energy estimate is similar to that one presented in Chapter 3, where we have constructed functions A^{ij} which are C^1 and close to a^{ij}. The closeness has been described in terms of the modulus of continuity ω. Since we now assume that the boundary ∂Q is $C^{1,\mathrm{Dini}}$, we can carry out the construction of the suitable smooth coefficients only locally. The proof of the fundamental result - the energy estimate - is more involved and is presented in detail.

10.1. Dini condition and preliminaries.

In this section we discuss the Dirichlet problem:

$$Lu = -D_i(a^{ij}(x)D_j u) = f(x) \quad \text{in} \quad Q, \tag{10.1}$$

$$u(x) = \phi(x) \quad \text{on} \quad \partial Q, \tag{10.2}$$

under weakened assumptions on a^{ij} and the boundary ∂Q. Our first objective is the derivation of the energy estimate.

Throughout this section, we assume that (A_1) of Section 3.1 (see Chapter 3) holds. The assumption (A_2) is weakened to the Dini continuity of $\{a^{ij}(x)\}$ on ∂Q, that is,

(A_2') There exists an increasing continuous function $\omega : [0,\infty) \to [0,\infty)$ with $\int_0^T \frac{dt}{\omega(t)} < \infty$ and $\omega(0) = 0$ such that

$$|a^{ij}(x) - a^{ij}(y)| \leqq \omega(|x-y|), \quad i,j = 1,...,n.$$

for all $x \in \partial Q$ and $y \in \bar{Q}$.

The boundary ∂Q is of class $C^{1,\mathrm{Dini}}$ and as usual we assume that $\phi \in L^2(\partial Q)$. The boundary condition (10.2) is understood in the following sense: to every $x^\circ \in \partial Q$ there is a neighbourhood V_{x° of x° in ∂Q such that

$$\lim_{\delta\to 0} \int_{V_{x^\circ}} [u(x+\delta\nu(x^\circ)) - \phi(x)]^2\, dS_x = 0, \tag{10.3}$$

where $\nu(x^\circ)$ is an inward normal at x° to ∂Q. This slightly differs from the approach adopted in Section 8.3 (see (8.14)), where the boundary data is recovered locally by approaching ∂Q through small surfaces parallel to ∂Q.

We first establish the energy estimate for solutions of the equation

$$Lu + \lambda u = f(x) \quad \text{in} \quad Q, \tag{10.4}$$

with the boundary condition (10.2), for large values of the parameter λ.

We need some notation and terminology. It follows from the regularity of ∂Q that for every $x^\circ \in \partial Q$ there exists $r_\circ > 0$ and $F_\circ \in C^1(R_{n-1})$ having the following properties:

$$F_\circ(0) = 0,\ DF_\circ(0) = 0,\ |DF_\circ(x')| \leqq \frac{1}{2} \text{ for all } x' \in R_{n-1}$$

and

$$Q \cap B(x^\circ, r_\circ) = B(x^\circ, r_\circ) \cap \{(x', x_n);\ x_n > F_\circ(x')\}.$$

Here (x', x_n) denotes coordinates of x in a local coordinate system with center at x° and the x_n–axis pointing in the direction of $\nu(x^\circ)$. Obviously, we have

$$\partial Q \cap B(x^\circ, r_\circ) = B(x^\circ, r_\circ) \cap \{(x', x_n);\ x_n = F_\circ(x')\}.$$

We may always assume that the neighbourhood appearing in (10.3) is such that $\partial Q \cap B(x^\circ, r_\circ) \subset V_{x^\circ}$ and therefore if $u \in \widetilde{W}^{1,2}(Q)$ is a solution of (10.4), (10.2), we may assume that

$$\lim_{\delta \to 0} \int_{B_{n-1}(0, \frac{2}{\sqrt{5}}) r_{x^\circ})} [u(x', F_\circ(x') + \delta) - u(x', F_\circ(x'))]^2\, dx' = 0, \tag{10.5}$$

where the subscript $n-1$ denotes that the ball is $(n-1)$–dimensional.

Let us set $l_\circ = \frac{r_\circ}{\sqrt{2}}$. Since the boundary is of $C^{1,\text{Dini}}$ class, we see that for all $x = (x', F_\circ(x'))$ and $z = (z', F_\circ(z'))$ in $B(x^\circ, r_\circ)$, we have

$$|DF_\circ(x') - DF_\circ(z')| \leqq \omega(|z' - x'|). \tag{10.5}$$

From the covering $\{B(x^\circ, l_\circ), x^\circ \in \partial Q\}$ of ∂Q, we choose a finite subcovering $\{B(x^m, l_m)\}$, $m = 1, ..., p$, with $l_m = \frac{r_m}{\sqrt{2}}$. We use the notation $B^m = B(x^m, l_m)$. A function from a local representation of ∂Q in B^m is denoted by F_m. We set

$$h = \frac{1}{3}\left(\frac{2}{\sqrt{5}} - \frac{\sqrt{2}}{2}\right) \min(r_1, ..., r_p)$$

$$\Pi_m^{l_m+h,h} = \{(x', x_n);\ |x'| < l_m + h,\ F_m(x') < x_n < F_m(x') + h\}.$$

It is obvious that $\Pi_m^{l_m+h,h} \subset B^m \cap Q$. We now choose $d_\circ = (0, \frac{h}{4})$ such that

$$Q^{3d_\circ} = \{x \in Q;\ r(x) \leqq 3d_\circ\} \subset \bigcup_{m=1}^{p} \Pi_m^{l_m,h}.$$

Setting $\Pi_m^h = \Pi_m^{l_m+d_\circ,h} \subset B^m \cap Q$. Finally, we set $Q'_m = Q_{3d_\circ} \cup \Pi_m^{l_m,h}$ and $Q_m = Q_{2d_\circ} \cup \Pi_m^h$. It is easy to see that

$$r(x) \leqq x_n - F_m(x') \leqq \frac{\sqrt{5}}{2} r(x') < \frac{4}{3} r(x) \tag{10.7}$$

for all $x \in \Pi_m^h$, $m = 1, \ldots, p$, and

$$2d_\circ \leqq x_n - F_m(x') < 3d_\circ \tag{10.8}$$

for all $x \in \partial Q^{2d_\circ} \cap \Pi_m^h$.

Let us now fix $x^m \in \partial Q$ and consider a local coordinate system with center at x^m. For brevity we write $F = F_m$. We define the mapping $\zeta : R_n \to R$ by $\zeta(x) = (x', x_n - F(x'))$ and let $\zeta^{-1}(y) = (y', y_n + F(y'))$. For a set E, we use the notation $\tilde{E} = \zeta(E)$.

We need the following technical lemmas.

Lemma 10.1. *Let $u \in W^{1,2}_{\text{loc}}(Q)$ be solution of (10.4) and let $\tilde{u}(y) = u(y', y_n + F(y'))$ and $\tilde{f}(y) = f(y', y_n + F(y'))$ for $y \in \tilde{Q}$. Then $\tilde{u}$ is a solution in $W^{1,2}_{\text{loc}}(\tilde{Q})$ of the equation*

$$-D_i(\tilde{a}^{ij}(y) D_j \tilde{u}) + \lambda \tilde{u} = \tilde{f}(y) \quad \text{in} \quad \tilde{Q},$$

where

$$\tilde{a}^{ij}(y) = a^{ij}(y', y_n + F(y')) \quad \text{for} \quad i,j = 1, \ldots, n-1,$$

$$\tilde{a}^{ni}(y) = \tilde{a}^{in}(y) = \tilde{a}^{ni}(y', y_n + F(y')) - \sum_{k=1}^{n-1} a^{ki}(y', y_n + F(y')) D_k F(y')$$

for $i = 1, \ldots, n-1$,

$$\begin{aligned} \tilde{a}^{nn}(y) = &\sum_{k,m=1}^{n-1} a^{km}(y', y_n + F(y')) D_k F(y') D_m F(y') \\ &- 2\sum_{k=1}^{n-1} a^{nk}(y', y_n + F(y')) D_k F(y') + a^{nn}(y', y_n + F(y')). \end{aligned}$$

Moreover

$$\gamma \leqq \gamma(1 + |DF(y')|^2) \leqq \tilde{a}^{nn}(y) \leqq \gamma_1(1 + |DF(y')|^2) \leqq \frac{5}{4}\gamma_1$$

where γ_1 is a constant depending on $\|a^{ij}\|_\infty$, $i,j = 1, \ldots, n$, and

$$|\tilde{a}^{ij}(y) - a^{ij}(z)| \leqq \tilde{\omega}(|y - z|)$$

for all $y \in \widetilde{\Pi}_m^m = \{y;\ |y'| \leqq l_m + d_\circ,\ 0 < y_m < h\}$ and all $z = (z', 0)$ with $|z'| < l_m + d_\circ$, where $\tilde{\omega}(t) = C\omega(\sqrt{2}t)$ and $C > 0$ depends on n and a^{ij}.

PROOF. We only show the Dini continuity of $\tilde{a}^{ij}$. First we observe that

$$\begin{aligned} |\zeta(x) - \zeta(z)|^2 &= |x' - z'|^2 + (x_n - F(x') - z_n + F(z'))^2 \\ &\leqq |x' - z'|^2 + (|x_n - z_n| + \frac{1}{2}|x' - z'|)^2 \leqq 2|x - z|^2 \end{aligned}$$

for all $x \in R_n$ and $z \in R_n$ and similarly

$$|\zeta^{-1}(y) - \zeta^{-1}(z)| \leqq \sqrt{2}|y - z|$$

for all $y \in R_n$ and $z \in R_n$. Hence

$$\begin{aligned}|\tilde{a}^{ij}(y) - \tilde{a}^{ij}(z)| &= |a^{ij}(\zeta^{-1}(y)) - a^{ij}(\zeta^{-1}(z))|\\ &\leqq \omega(|\zeta^{-1}(y) - \zeta^{-1}(z)|) \leqq \omega(\sqrt{2}|y - z|),\ i,j = 1,...,n-1,\end{aligned}$$

and

$$\begin{aligned}&|\tilde{a}^{in} - \tilde{a}^{in}(z)| \leqq \sum_{k=1}^{n-1} |\tilde{a}^{ik}(y)D_kF(y') - \tilde{a}^{ik}(z)D_kF(z')|\\ &+ \tilde{a}^{in}(y', y_n + F(y')) - \tilde{a}^{in}(z', z_n + F(z'))|\\ &\leqq \sum_{k=1}^{n-1}\Big[|\tilde{a}^{ik}(y)||D_kF(y') - D_kF(z')| + +|D_kF(z')||\tilde{a}^{ik}(y) - \tilde{a}^{ik}(z)|\Big]\\ &+ \omega(|\zeta^{-1}(y) - \zeta^{-1}(z)) \leqq \sum_{k=1}^{n-1} \|a^{ik}\|_\infty |D_kF(y') - D_kF(z')|\\ &+ \frac{n-1}{2}\omega(\sqrt{2}|y - z|) + \omega(\sqrt{2}|y - z|),\end{aligned}$$

$i = 1, ..., n-1$. Similarly we obtain the estimate

$$|\tilde{a}^{nn}(y) - \tilde{a}^{nn}(z)| \leqq C\omega(\sqrt{2}|y - z|)$$

and the result follows from (10.6).

We now state an obvious result.

Lemma 10.3. *Let* $\{a^{ij}(y)\}$, $i,j = 1,...,n$, *be a matrix defined on* $\widetilde{\Pi}^h_m$ *by*

$$a^{ij}_\circ(y) = \tilde{a}^{ij}(y), \quad i,j = 1,...,n-1,$$

$$a^{in}_\circ(y) = a^{ni}_\circ(y) = \frac{1}{|B_{n-1}(0, y_n)|}\int_{B_{n-1}(y', y_n)} \tilde{a}^{in}(\xi', 0)\, d\xi', \quad i = 1, ..., n-1$$

and $\tilde{a}^{nn}_\circ(y) = \tilde{a}^{nn}(y', 0)$. *Then*

$$|a^{nn}_\circ(y) - \tilde{a}^{nn}(y)| \leqq \tilde{\omega}(y_n),$$

$$|a^{in}_\circ(y) - \tilde{a}^{in}(y)| \leqq \frac{1}{|B_{n-1}(0, y_n)|}\int_{B_{n-1}(y', y_n)} |\tilde{a}^{in}(\xi', 0) - \tilde{a}^{in}(y', y_n)|\, d\xi \leqq \tilde{\omega}(\sqrt{2}y_n)$$

and

$$|D_i a^{in}_\circ(y', y_n)| \leqq \frac{1}{|B_{n-1}(0, y_n)|}\int_{B_{n-2}(\hat{\xi}', y_n)} \tilde{\omega}(2y_n)\, d\hat{\xi}' = \frac{n-1}{y_n}\tilde{\omega}(2y_n),$$

for $i = 1, ..., n-1$, where $\hat{\xi}'$ stands for ξ' with ξ_i omitted and $\tilde{\omega}(t) = C\omega(2\sqrt{2}t)$ with a constant $C > 0$ depending on n and $\|a^{ij}\|_\infty$.

It follows from the above estimates that

$$\left[\sum_{i=1}^{n} |a_\circ^{in}(y) - \tilde{a}^{in}(y)|^2\right]^{\frac{1}{2}} \leqq \tilde{\omega}(y_n) \tag{10.10}$$

and

$$|D_i a_\circ^{in}(y)| \leqq \frac{\tilde{\omega}(y_n)}{y_n}, \quad i = 1, ..., n-1. \tag{10.11}$$

We also need the following version of the Caccioppoli inequality (see Lemma 5.1)

$$\int_K |Du(x)|^2\, dx \leqq C_\circ \left[\frac{1}{\sigma^2} \int_Q u(x)^2\, dx + \sigma^2 \int_Q f(x)^2\, dx\right],$$

where K is a compact subset of Q and $\sigma = \operatorname{dist}(K, \partial Q)$ and $C_\circ > 0$ is a constant independent of u and σ.

10.2. Energy estimate.

We are now in a position to establish the energy estimate.

Theorem 10.1.. *Let $\phi \in L^2(\partial Q)$. There exist positive constants $\lambda_\circ$, d and C such that any solution $u \in W^{1,2}_{\text{loc}}(Q)$ of (10.4), (10.2), with $\lambda \leqq \lambda_\circ$, satisfies*

(10.13)

$$\int_Q |Du(x)|^2 r(x)\, dx + (\lambda - \lambda_\circ) \int_Q u(x)^2 r(x)\, dx + \sup_{0<\delta<d} \int_{\partial Q_\delta} u(x)^2\, dS_x$$
$$\leqq C\left[\int_{\partial Q} \phi(x)^2\, dS_x + \int_Q f(x)^2 r(x)^\theta\, dx\right].$$

PROOF. We set $\tilde{u}(y) = u(y', y_n + F(y'))$, then by Lemma 10.1 $\tilde{u} \in W^{1,2}_{\text{loc}}(\tilde{Q})$ and satisfies (10.9). We fix a positive number $\delta_\circ \leqq \frac{d_\circ}{2}$ and in the sequel we shall impose further conditions on $\delta_\circ$. For $\delta \in (0, \delta_\circ)$ we define a function $\rho_\delta : \tilde{Q}_m \to [0, \infty)$ in the following way

$$\rho_\delta(y) = \begin{cases} 0 & \text{for } |y'| < l_m + d_\circ,\ 0 < y_n < \delta, \\ y_n - \delta & \text{for } |y'| < l_m + d_\circ,\ \delta < y_n < 4\delta_\circ, \\ 4\delta_\circ - \delta & \text{elsewhere in } \tilde{Q}_m. \end{cases}$$

It follows from (10.8) that $\rho_\delta \in W^{1,\infty}(\tilde{Q}_m)$ and $\|D\rho_\delta\|_\infty \leqq 1$. Moreover, $3\delta_\circ \leqq 4\delta_\circ - \delta = \rho_\delta(\zeta(x)) \leqq 4\delta_\circ$ for $x \in Q_{2d_\circ}$ and by (10.7)

$$\min\{3\delta_\circ, \max\{0, r(x) - \delta\}\} \leqq \rho_\delta(\zeta(x)) \leqq \min\{4\delta_\circ, \max\{0, \frac{4}{3} r(x) - \delta\}\}$$
$$= \frac{4}{3} \min\{3\delta_\circ, \max\{0, r(x) - \frac{3}{4}\delta\}\},$$

for $x \in \Pi_m^h$. Hence

$$r_\delta(x) \leqq \rho_\delta(\zeta(x)) \leqq \frac{4}{3} r_{\frac{3}{4}\delta}(x) \tag{10.14}$$

for $x \in Q_m$, where $r_\delta(x) = \min\{3\delta_\circ, \max\{0, r(x) - \delta\}\}$.
Let $\Psi \in C^1(\bar{Q})$ be such that $\Psi(x) = 1$ for $x \in Q'_m$, $\Psi(x) = 0$ for $x \in Q^{\frac{5}{2}d_\circ} - \Pi_m^{l_m + \frac{1}{2}d_\circ, h}$ and $0 \leqq \Psi \leqq 1$ on $\bar{Q}$. We also assume that the function $\widetilde{\Psi}(y) = \Psi(\zeta^{-1}(y))$ is independent of y_n. It is clear that $\widetilde{\Psi}$ satisfies the following conditions:

$$\widetilde{\Psi}(y) = \widetilde{\Psi}(y', y_n) = 1 \ \text{ for } \ |y'| < l_m \ \text{ and } \ 0 < y_n < h,$$

$$\widetilde{\Psi}(y) = \widetilde{\Psi}(y', y_n) = 0 \ \text{ for } \ l_m + \frac{1}{2}d_\circ < |y'| < l_m + d_\circ \ \text{ and } \ 0 < y_n < 2d_\circ.$$

Taking $\tilde{\eta}(y) = \rho_\delta(y)\widetilde{\Psi}(y)\tilde{u}(y)$, with $\tilde{\eta}(y) = 0$ for $y \in \tilde{Q} - \tilde{Q}_m$ as a test function in (10.9) we obtain on substitution

(10.15)
$$\begin{aligned}
\tilde{I}_1^m &= \int_{\tilde{Q}} \rho_\delta \widetilde{\Psi} \tilde{u} \tilde{f}\, dy = \int_{\tilde{Q}_m} \tilde{\eta} \tilde{f}\, dy = \int_{\tilde{Q}_m} [\tilde{a}^{ij} D_i \tilde{\eta} D_j \tilde{u} + \lambda \tilde{\eta} \tilde{u}]\, dy \\
&= \int_{\tilde{Q}_m} [\lambda \rho_\delta \widetilde{\Psi} \tilde{u}^2 + \rho_\delta \widetilde{\Psi} \tilde{a}^{ij} D_i \tilde{u} D_j \tilde{u} + \rho_\delta \tilde{u} \tilde{a}^{ij} D_i \widetilde{\Psi} D_j \tilde{u} + \tilde{u} \widetilde{\Psi} \tilde{a}^{ij} D_i \rho_\delta D_j \tilde{u}]\, dy \\
&= \int_{\tilde{Q}_m} \lambda \rho_\delta \widetilde{\Psi} \tilde{u}^2\, dy + \int_{\tilde{Q}_m} \rho_\delta \widetilde{\Psi} \tilde{a}^{ij} D_i \tilde{u} D_j \tilde{u}\, dy + \int_{\tilde{Q}_m} \rho_\delta \tilde{u} \tilde{a}^{ij} D_i \widetilde{\Psi} D_j \tilde{u}\, dy \\
&+ \int_\delta^{4\delta_\circ} dy_n \int_{|y'| < l_m + d_\circ} \tilde{u} \widetilde{\Psi} (\tilde{a}^{ij} - \tilde{a}_\circ^{ij}) D_i \rho_\delta D_j \tilde{u}\, dy' \\
&- \frac{1}{2} \int_\delta^{4\delta_\circ} dy_n \int_{|y'| < l_m + d_\circ} \widetilde{\Psi} \tilde{u}^2 \sum_{i=1}^{n-1} D_i a_\circ^{in}\, dy' \\
&- \frac{1}{2} \int_\delta^{4\delta_\circ} dy_n \int_{|y'| < l_m + d_\circ} \tilde{u}^2 \sum_{i=1}^{n-1} a_\circ^{in} D_i \widetilde{\Psi}\, dy' \\
&+ \frac{1}{2} \int_{|y'| < l_m + d_\circ} \tilde{a}^{nn}(y', 0) \widetilde{\Psi}(y') \tilde{u}(y', 4\delta_\circ)^2\, dy' \\
&- \frac{1}{2} \int_{|y'| < l_m + d_\circ} \tilde{a}^{nn}(y', 0) \widetilde{\Psi}(y') \tilde{u}(y', \delta)^2\, dy' \\
&= \widetilde{J}_1^m + \widetilde{J}_2^m + \widetilde{I}_2^m + \widetilde{I}_3^m + \widetilde{I}_4^m + \widetilde{I}_5^m + \widetilde{I}_6^m - \widetilde{J}_3^m.
\end{aligned}$$

We estimate the integrals $\widetilde{J}_i^m$ from below and the integrals $\widetilde{I}_i^m$ from above. By (10.14) we have

$$\widetilde{J}_1^m = \widetilde{J}_1^m(\delta) = \lambda \int_{\tilde{Q}_m} \rho_\delta(y) \widetilde{\Psi}(y) \tilde{u}(y)^2\, dy \geqq \lambda \int_{Q'_m} r_\delta(x) u(x)^2\, dx = \lambda J_1^m(\delta),$$

$$\widetilde{J}_2^m = \widetilde{J}_2^m(\delta) = \int_{\tilde{Q}_m} \rho_\delta \widetilde{\Psi}(y)\tilde{a}^{ij}(y)D_i\tilde{u}(y)D_j\tilde{u}(y)\,dy$$

$$\geqq \int_{Q'_m} r_\delta(x)a^{ij}(x)D_iu(x)D_ju(x)\,dx \geqq \gamma \int_{Q'_m} r_\delta(x)|Du(x)|^2\,dx = J_2^m(\delta)$$

and

$$\widetilde{J}_3^m = \widetilde{J}_3^m(\delta) = \frac{1}{2}\int_{|y'|<l_m+d_\circ} \tilde{a}^{nn}(y',0)\widetilde{\Psi}(y')u(y',\delta)^2\,dy'$$

$$\geqq \frac{\gamma}{2}\int_{|y'|<l_m+d_\circ} \widetilde{\Psi}(y')\tilde{u}(y',\delta)^2\,dy' = J_3^m(\delta).$$

Let us set

$$M(\delta) = \max_{\delta\leqq y_n\leqq\delta_\circ}\int_{|y'|<l_m} \widetilde{\Psi}(y')u(y',y_n)^2\,dy',$$

we then have

$$|\widetilde{I}_1^m| = |\int_{\tilde{Q}_m} \rho_\delta\widetilde{\Psi}\tilde{u}f\,dy|$$

$$\leqq \int_\delta^{\delta_\circ} dy_n \int_{|y'|<l_m+d_\circ} (y_n-\delta)\widetilde{\Psi}(y')|\tilde{u}(y',y_n)\tilde{f}(y',y_n)|\,dy'dy_n$$

$$+4\delta_\circ\int_{Q_m-\Pi_m^{l_m+d_\circ,\delta_\circ}} \Psi(x)|u(x)f(x)|\,dx$$

$$\leqq M(\delta)^{\frac{1}{2}}\left[\int_\delta^{\delta_\circ} y_n^{2-\theta}\,dy_n\right]^{\frac{1}{2}}\left[\int_\delta^{\delta_\circ} dy_n\int_{|y'|<l_m+d_\circ} y_n^\theta|\tilde{f}(y',y_n)|^2\,dy'\right]^{\frac{1}{2}}$$

$$+\frac{C_1'}{\delta_\circ^{\frac{\theta-1}{2}}}\int_{Q_m-\Pi_m^{l_m+d_\circ,\delta_\circ}} r(x)^{\frac{\theta+1}{2}}|u(x)f(x)|\,dx \leqq \epsilon_1 M(\delta)$$

$$+C_1\left[\frac{\delta_\circ^{3-\theta}}{\epsilon_1}\int_Q r(x)^\theta f(x)^2\,dx + \frac{1}{\delta_\circ^2}\int_Q r(x)u(x)^2\,dx\right] = \epsilon_1 M(\delta) + I_1^m(\delta_\circ,\epsilon_1),$$

where

$$I_1^m(\delta_\circ,\epsilon_1) = C_1\left[\frac{\delta_\circ^{3-\theta}}{\epsilon_1}\int_Q r(x)^\theta f(x)^2\,dx + \frac{1}{\delta_\circ^2}\int_Q r(x)u(x)^2\,dx\right],$$

and C_1' and C_2 are constants independent of $\delta_\circ$ and the constant ϵ_1 will be chosen later. Similarly

$$|\widetilde{I}_2^m| = |\int_{\tilde{Q}_m} \rho_\delta(y)\tilde{u}(y)\tilde{a}^{ij}(y)D_i\widetilde{\Psi}(y)D_j\tilde{u}(y)\,dy$$

$$\leqq \frac{4}{3}\int_{Q_m} r_{\frac{3}{4}\delta}(x)|u(x)a^{ij}(x)D_i\Psi(x)D_ju(x)|\,dx$$

$$\leqq \frac{4}{3}C_2\|\Psi\|_{C^1(\bar{Q})}\left[\int_{Q_m} r_{\frac{3}{4}\delta}(x)u(x)^2\,dx\right]^{\frac{1}{2}}\left[\int_{Q-Q^{\frac{5}{2}d_\circ}\cup\Pi_m^{l_m+\frac{d_\circ}{2},h}} r_{\frac{3}{4}\delta}(x)|Du(x)|^2\,dx\right]^{\frac{1}{2}}$$

$$= I_2^{m'}(\delta),$$

where $C_2 >$ depends on $\|a^{ij}\|_{\infty}$. Using the estimate (10.12) and the fact that $I_2^{m'}(\delta)$ is increasing as $\delta \to 0$, we obtain

$$I_2^{m'}(\delta) = \frac{4C_2}{3}\|\Psi\|_{C^1(\bar{Q})}\left[\int_Q r(x)u(x)^2\,dx\right]^{\frac{1}{2}}\left[\frac{\delta}{2}\int_{\Pi_m^{l_m+x\frac{d_o}{2},h}\cap Q^{\frac{5\delta}{4}}-Q^{\frac{3\delta}{4}}}|Du(x)|^2\,dx\right.$$

$$\left.+2\int_Q r_\delta(x)|Du(x)|^2\,dx\right]^{\frac{1}{2}}$$

$$\leqq \epsilon_2\int_Q r_\delta(x)|Du(x)|^2\,dx + C_2'\left[\frac{1}{\epsilon_2}r(x)u(x)^2\,dx + \delta^{3-\theta}\int_Q r(x)^\theta f(x)^2\,dx\right.$$

$$\left.+\max_{\frac{\delta}{2}\leqq y_n\frac{3}{\delta}}\int_{|y'|<l_m+\delta_o}u(y',y_n)^2\,dy'\right] = \epsilon_2\int_Q r_\delta(x)|Du(x)|^2\,dx + I_2^m(\delta,\epsilon_2),$$

where $\epsilon_2 \in (0,1)$ and

$$I_2^m(\delta,\epsilon_2) = C_2'\left[\frac{1}{\epsilon_2}\int_Q r(x)u(x)^2\,dx + \delta^{3-\theta}f(x)^2\,dx\right.$$

$$\left.+\max_{\frac{\delta}{2}\leqq y_n 3\delta}\int_{|y'|<l_m+d_o}\tilde{u}(y',y_n)^2\,dy'\right]$$

for some constant $C_2' > 0$. We now impose the first condition on δ_o, that is,

$$\tilde{\omega}(4\delta_o) \leqq 1, \tag{10.16}$$

where $\tilde{\omega}$ is the function appearing in (10.10) and (10.11). To estimate $\widetilde{I_3^m}$ we use (10.10) and (10.11) to obtain

$$|\widetilde{I_3^m}| = |\int_\delta^{4\delta_o}dy_n\int_{|y'|<l_m+d_o}\tilde{u}(y',y_n)\widetilde{\Psi}(y)(\tilde{a}^{ij}(y)-\tilde{a}_o^{ij}(y))D_i\rho_\delta(y_n)D_j\tilde{u}(y',y_n)\,dy'$$

$$\leqq \int_\delta^{4\delta_o}dy_n\int_{|y'|<l_m+d_o}|\tilde{u}(y)|\widetilde{\Psi}(y')\sum_{i=1}^n|\tilde{a}^{in}(y)-a_o^{in}(y)||D_i\tilde{u}(y)|\,dy'$$

$$\leqq \int_\delta^{4\delta_o}dy_n\int_{|y'|<l_m+d_o}|\tilde{u}(y)|\widetilde{\Psi}(y')|D\tilde{u}(y)|\tilde{\omega}(y_n)\,dy'$$

$$\leqq \left[\int_\delta^{\delta_o}dy_n\int_{|y'|<l_m+d_o}\widetilde{\Psi}(y')y_n|D\tilde{u}(y)|^2\,dy'\right]^{\frac{1}{2}}M(\delta)^{\frac{1}{2}}\left[\int_0^{\delta_o}\frac{\tilde{\omega}(y_n)^2}{y_n}\,dy_n\right]^{\frac{1}{2}}$$

$$+\tilde{\omega}(4\delta_o)\int_{\delta_o}^{4\delta_o}dy_n\int_{|y'|<l_m+d_o}|\tilde{u}(y)||D\tilde{u}(y)|\,dy' \leqq \frac{\epsilon_2}{4}\int_{\Pi_m^{l_m+\frac{d_o}{2},h}}r_{\frac{3}{4}\delta}(x)|Du(x)|^2\,dx$$

$$+\frac{8}{\epsilon_2}\omega_1(\delta_o)M(\delta) + \frac{\epsilon_2}{4}\int_{\Pi_m^{l_m+\frac{d_o}{2},h}}r_{\frac{3}{4}\delta_o}(x)|Du(x)|^2\,dx$$

$$+\frac{C_3'}{\epsilon_o\delta_o^2}\int_{Q_m}r_{\frac{3}{4}\delta_o}(x)u(x)^2\,dx \leqq \frac{\epsilon_2}{2}\int_{\Pi_m^{l_m+\frac{d_o}{2},h}}r_{\frac{3}{4}\delta}(x)|Du(x)|^2\,dx$$

$$+\frac{C_3'}{\epsilon_2\delta_o^2}\int_Q r(x)u(x)^2\,dx + \frac{8}{\epsilon_2}\omega_1(\delta_o)M(\delta) = I_3^{m'}(\delta,\delta_o,\epsilon_2)$$

$$+\frac{8}{\epsilon_2}\omega_1(\delta_o)M(\delta),$$

where

$$\omega_1(\delta_\circ) = \int_0^{\delta_\circ} \frac{\tilde{\omega}(t)}{t}\, dt \to 0, \text{ as } \delta_\circ \to 0,$$

and $C_3' > 0$ is a constant independent of δ, $\delta_\circ$ and u. It follows from the Caccioppoli inequality (10.12) that

$$I_3^{m'}(\delta,\delta_\circ,\epsilon_2) = \frac{\epsilon_2}{2}\int_{\Pi_m^{l_m+\frac{d_0}{2},h}} r_{\frac{3}{4}\delta}(x)|Du(x)|^2\, dx + \frac{C_3'}{\epsilon_2\delta_\circ^2}\int_Q r(x)u(x)^2\, dx$$

$$\leqq \frac{\epsilon_2}{2}\left[\frac{\delta}{2}\int_{\Pi_m^{l_m+\frac{d_0}{2},h}\cap Q^{\frac{5}{4}\delta}-Q^{\frac{3}{4}\delta}} |Du(x)|^2\, dx + 2\int_Q r_\delta(x)|Du(x)|^2\, dx\right]$$

$$+ \frac{C_3'}{\epsilon_2\delta_\circ^2}\int_Q r(x)u(x)^2\, dx$$

$$\leqq \epsilon_2\int_Q r_\delta(x)|Du(x)|^2\, dx + C_3\left[\epsilon_2 \max_{\frac{\delta}{2}\leqq y_n\leqq 3\delta}\int_{|y'|<l_m+d_\circ} u(y',y_n)^2\, dy'\right.$$

$$\left. + \epsilon_2\delta^{3-\theta}\int_Q r(x)^\theta f(x)^2\, dx + \frac{1}{\epsilon_2\delta_\circ^2}\int_Q r(x)u(x)^2\, dx\right]$$

$$\leqq \epsilon_2\int_Q r_\delta(x)|Du(x)|^2\, dx + I_3^m(\delta,\delta_\circ,\epsilon_2),$$

where

$$I_3^m = C_3\left[\max_{\frac{\delta}{2}\leqq y_n\leqq 3\delta}\int_{|y'|<l_m+d_\circ} u(y',y_n)^2\, dy'\right.$$

$$\left. + \delta^{3-\theta}\int_Q r(x)^\theta f(x)^2\, dx + \frac{1}{\epsilon_2\delta_\circ^2}\int_Q r(x)u(x)^2\, dx\right].$$

Using (10.11) and (10.16), we find that

$$|\tilde{I}_4^m| = \frac{1}{2}\int_\delta^{4\delta_\circ} dy_n \int_{|y'|<l_m+d_\circ} \tilde{\Psi}(y')u(y',y_n)^2 \sum_{i=1}^{n-1} |D_i a_\circ^{in}(y',y_n)|\, dy'$$

$$\leqq \frac{1}{2}M(\delta)(n-1)\int_0^{\delta_\circ} \frac{\tilde{\omega}(y_n)}{y_n}\, dy_n + \frac{n-1}{2}\int_{\delta_\circ}^{4\delta_\circ} dy_n \int_{|y'|<l_m+d_\circ} u(y)^2\frac{\omega(y_n)}{y_n}\, dy'$$

$$\leqq \frac{n-1}{2}\omega_1(\delta_\circ)M(\delta) + \frac{C_4}{\delta_\circ^2}\int_Q u(x)^2 r(x)\, dx = \frac{n-1}{2}\omega_1(\delta_\circ)M(\delta) + I_4^m(\delta_\circ)$$

and

$$|\tilde{I}_5^m| \leqq \frac{1}{2}\int_\delta^{4\delta_\circ} dy_n \int_{|y'|<l_m+d_\circ} \tilde{u}(y)^2 \sum_{i=1}^{n-1} |a_\circ^{in}(y)D_i\tilde{\Psi}(y)|\, dy_n \leqq C_5\int_Q u(x)^2\, dx = I_5^m.$$

Finally, by virtue of (10.12) and the trace estimate, we have

$$
\begin{aligned}
|I_6^m| &= \frac{1}{2}\Big|\int_{|y'|<l_m+d_\circ} \tilde{a}^{nn}(y,0)\widetilde{\Psi}(y')\tilde{u}(y',4\delta_\circ)^2\,dy'\Big| \\
&\leqq \frac{5}{8}\gamma_1\Big[\int_{|y'|<l_m+d_\circ} \tilde{u}(y',3d_\circ)^2\,dy' + 2\int_{4\delta_\circ}^{3d_\circ} dy_n \int_{|y'|<l_m+\frac{d_\circ}{2}} |\tilde{u}(y)||D\tilde{u}(y)|\,dy'\Big] \\
&\leqq C_6'\Big[\|u\|_{W^{1,2}(Q-Q^{\frac{5}{2}d_\circ})} + \frac{1}{\delta_\circ}\int_{3\delta_\circ}^{4d_\circ} dy_n \int_{|y'|<l_m+d_\circ} \tilde{u}(y',y_n)^2\,dy' \\
&+ \delta_\circ^3 \int_{3\delta_\circ}^{4d_\circ} dy_n \int_{|y'|<l_m+d_\circ} \tilde{f}(y)^2\,dy'\Big] \leqq C_6''\Big[\int_{Q-Q^{2d_\circ}} u(x)^2\,dx \\
&+ \frac{1}{\delta_\circ^2}\int_Q u(x)^2 r(x)\,dx + d_\circ^{3-\theta}\int_Q f(x)^2 r(x)^\theta\,dx\Big] \\
&\leqq C_6\Big[\frac{1}{\delta_\circ^2}\int_Q u(x)^2 r(x)\,dx + \int_Q f(x)^2 r(x)^\theta\,dx\Big] = I_6^m(\delta_\circ).
\end{aligned}
$$

Inserting the estimates for $\widetilde{I}_i^m$ and $\widetilde{J}_i^m$ into (10.15), we obtain the estimate for $M(\delta)$:

(10.17)

$$
\begin{aligned}
M(\delta) &= \max_{\delta\leqq y_n\leqq\delta_\circ} \int_{|y'|<l_m+d_\circ} \widetilde{\Psi}(y')\tilde{u}(y',y_n)^2\,dy' \leqq \frac{2}{\gamma}\max_{\delta\leqq y_n\leqq\delta_\circ} \widetilde{J}_3^m \\
&\leqq \frac{2}{\gamma}\Big[\widetilde{J}_1^m(\delta) + \widetilde{J}_2^m(\delta) + \epsilon_1 M(\delta) + I_1^m(\delta_\circ,\epsilon_1) + I_2^{m'}(\delta) \\
&+ \Big(\frac{8}{\epsilon_2} + \frac{n-1}{2}\Big)\omega_1(\delta_\circ)M(\delta) + I_3^{m'}(\delta,\delta_\circ,\epsilon_2) + I_4^m(\delta_\circ) + I_5^m + I_6^m(\delta_\circ)\Big].
\end{aligned}
$$

We now choose $\epsilon_1 \leqq \frac{\gamma}{8}$ and impose the second condition on $\delta_\circ$ (see (10.16)):

$$
\Big(\frac{8}{\epsilon_2} + \frac{n-1}{2}\Big)\omega_1(\delta_\circ) \leqq \epsilon_1. \tag{10.18}
$$

Consequently, using the estimates for $I_2^{m'}$ and $I_3^{m'}$ in terms of I_2^m, I_3^m and $\int_Q |Du(x)|^2 r_\delta(x)\,dx$, we obtain

(10.19)

$$
\begin{aligned}
M(\delta) &\leqq \frac{4}{\gamma}\Big[\widetilde{J}_1^m(\delta) + \widetilde{J}_2^m(\delta) + I_1^m(\delta_\circ,\epsilon_1) + I_2^{m'}(\delta) + I_3^{m'}(\delta,\delta_\circ,\epsilon_2) \\
&+ I_4^m(\delta_\circ) + I_5^m + I_6^m(\delta_\circ)\Big] \\
&\leqq \frac{4}{\gamma}\Big[\widetilde{J}_1^m(\delta) + \widetilde{J}_2^m(\delta) + I_1^m(\delta_\circ,\epsilon_1) + I_2^m(\delta,\epsilon_2) + I_3^m(\delta_1,\delta_\circ,\epsilon_2) \\
&+ 2\epsilon_2\int_Q |Du(x)|^2 r_\delta(x)\,dx + I_4^m(\delta_\circ) + I_5^m + I_6^m(\delta_\circ)\Big].
\end{aligned}
$$

On the other hand we can calculate $\widetilde{J}_1^m + \widetilde{J}_2^m$ for the identity (10.15) and then using the estimates for I_i^m $(i = 1, ..., 6)$ we obtain

$$\widetilde{J}_1^m(\delta) + \widetilde{J}_2^m(\delta) \leqq \widetilde{J}_3^m(\delta) + 2\epsilon_1 M(\delta) + I_1^m(\delta_\circ, \epsilon_1) + I_2^m(\delta, \epsilon_2)$$
$$+ I_3^m(\delta, \delta_\circ, \epsilon_2) + 2\epsilon_2 \int_Q |Du(x)|^2 r_\delta(x)\, dx + I_4^m(\delta_\circ) + I_5^m + I_6^m(\delta_\circ)$$

and hence by (10.19) we get

$$\widetilde{J}_1^m(\delta)+\widetilde{J}_2^m(\delta) \leqq \widetilde{J}_3^m(\delta) + \frac{8\epsilon_1}{\gamma}\Big[\widetilde{J}_1^m(\delta) + \widetilde{J}_2^m(\delta)\Big]$$
$$+ 2\epsilon_2 (1 + \frac{8\epsilon_1}{\gamma}) \int_Q |Du(x)|^2 r_\delta(x)\, dx + (1 + \frac{8\epsilon_1}{\gamma})\Big[I_1^m(\delta_\circ, \epsilon_1)$$
$$+ I_2^m(\delta, \epsilon_2) + I_3^m(\delta, \delta_\circ, \epsilon_2) + I_4^m(\delta_\circ) + I_6^m(\delta_\circ)\Big].$$

Setting

$$\epsilon_1 = \frac{\gamma}{16} \quad \text{and} \quad \epsilon_2 = \frac{\gamma}{12p},$$

the condition (10.18) becomes

$$(10.16) \qquad (\frac{96p}{\gamma} + \frac{n-1}{2})\omega_1(\delta_\circ) \leqq \frac{\gamma}{16}.$$

With this choice of ϵ_1 we get

$$\widetilde{J}_1^m(\delta) + \widetilde{J}_2^m(\delta) \leqq 2\widetilde{J}_3^m + \frac{\delta}{2p}\int_Q |Du(x)|^2 r_\delta(x)\, dx$$
$$+ 3\Big[I_1^m(\delta_\circ) + I_2^m(\delta) + I_3^m(\delta, \delta_\circ) + I_4^m(\delta_\circ) + I_5^m + I_6^m(\delta_\circ)\Big].$$

Since $Q^{3d_\circ} \subset \bigcup_{m=1}^p \Pi_m^{l_m, h}$, we obtain from the last inequality, summing in p, that

$$\lambda \int_Q u(x)^2 r_\delta(x)\, dx + \gamma \int_Q |Du(x)|^2 r_\delta(x)\, dx$$
$$\leqq \sum_{m=1}^p (\widetilde{J}_1^m(\delta) + \widetilde{J}_2^m(\delta)) \leqq 2 \sum_{m=1}^p \widetilde{J}_3^m(\delta) + \frac{\gamma}{2}\int_Q |Du(x)|^2 r_\delta(x)\, dx$$
$$+ 3\sum_{m=1}^p \Big[I_1^m(\delta_\circ) + I_2^m(\delta) + I_3^m(\delta, \delta_\circ) + I_4^m(\delta_\circ) + I_5^m + I_6^m(\delta_\circ)\Big]$$

and consequently

$$(10.21) \qquad \frac{2\lambda}{\gamma}\int_Q u(x)^2 r_\delta(x)\, dx + \int_Q |Du(x)|^2 r_\delta(x)\, dx$$
$$\leqq \frac{6}{\gamma}\sum_{m=1}^p \Big[\widetilde{J}_3^m(\delta) + I_1^m(\delta_\circ) + ... + I_6^m(\delta_\circ)\Big].$$

Since the right hand side of (10.21) is bounded function for $\delta \in (0,\delta_\circ)$ and $\int_Q |Du(x)|^2 r(x)\,dx < \infty$, letting $\delta \to 0$, we obtain

$$(10.22)\quad \frac{2\lambda}{\gamma}\int_Q u(x)^2 r(x)\,dx + \int_Q |Du(x)|^2 r(x)\,dx$$
$$\leqq C_7\Big[\int_{\partial Q}\phi(x)^2\,dS_x + \int_Q f(x)^2 r(x)^\theta\,dx + \frac{1}{\delta_\circ^2}\int_Q u(x)^2 r(x)\,dx + \int_Q u(x)^2\,dx\Big].$$

Now the proof is similar to that of Theorem 3.5 (see Chapter 3); that is, we need the estimate of $\int_Q u(x)^2\,dx$. By virtue of (10.17), we have

$$(10.23)\quad \int_0^{\delta_\circ} dy_n \int_{|y'|<l_m} u(y',y_n)^2\,dy'dy_n \leqq \int_0^{\delta_\circ} M(y_n)\,dy_n \leqq M(0)\delta_\circ$$
$$\leqq C_8\delta_\circ\Big[(\lambda+\frac{1}{\delta_\circ^2})\int_Q u(x)^2 r(x)\,dx + \int_Q |Du(x)|^2 r(x)\,dx$$
$$+ \int_Q f(x)^2 r(x)^\theta\,dx + \int_Q u(x)^2\,dx + \int_{\partial Q}\phi(x)^2\,dS_x\Big]$$

and summing over m we arrive at

$$\int_Q u(x)^2\,dx \leqq C_9'\Big[\sum_{m=1}^{p}\int_{\Pi_m^{l_m,\delta_\circ}} u(x)^2\,dx + \frac{1}{\delta_\circ}\int_Q u(x)^2 r(x)\,dx\Big]$$
$$\leqq C_9\delta_\circ\Big[(\lambda+\frac{1}{\delta_\circ^2})\int_Q u(x)^2 r(x)\,dx + \int_Q |Du(x)|^2 r(x)\,dx$$
$$+ \int_Q f(x)^2 r(x)^\theta\,dx + \int_Q u(x)^2\,dx + \int_{\partial Q}\phi(x)^2\,dS_x\Big].$$

Combining this with (10.23), we get

$$\int_Q u(x)^2\,dx + \int_Q |Du(x)|^2 r(x)\,dx + \frac{2\lambda}{\gamma}\int_Q u(x)^2 r(x)\,dx$$
$$\leqq \Big[C_9\delta_\circ(\lambda+C_7)(1+\frac{1}{\delta_\circ^2}) + \frac{C_7}{\delta_\circ^2}\Big]\int_Q u(x)^2 r(x)\,dx$$
$$+ C_9\delta_\circ(1+C_7)\int_Q |Du(x)|^2 r(x)\,dx$$
$$+ (C_7+(1+C_7)C_9\delta_\circ)\Big[\int_Q f(x)^2 r(x)^\theta\,dx + \int_{\partial Q}\phi(x)^2\,dS_x\Big].$$

Imposing now the third condition on $\delta_\circ$, that is,

$$\delta_\circ C_9(1+C_7) \leqq \min\big(\frac{1}{\gamma},\frac{1}{2}\big),$$

we get

$$(10.24)\quad \int_Q u(x)^2\,dx + \int_Q |Du(x)|^2 r(x)\,dx + \frac{2\lambda}{\gamma}\int_Q u(x)^2 r(x)\,dx$$
$$\leqq 2\Big[C_9(1+C_7)\delta_\circ + \frac{C_7}{\delta_\circ^2}\Big]\int_Q u(x)^2 r(x)\,dx$$
$$+2(C_7+(1+C_7)C_9\delta_\circ)\Big[\int_Q f(x)^2 r(x)^\theta\,dx + \int_{\partial Q}\phi(x)^2\,dS_x\Big]$$
$$\leqq \frac{C_{10}}{\delta_\circ}\int_Q u(x)^2 r(x)\,dx + C_{11}\Big[\int_Q f(x)^2 r(x)^\theta\,dx + \int_{\partial Q}\phi(x)^2\,dS_x\Big].$$

The estimate (10.23) also implies that

$$M(0) \leqq C_8\Big[(\lambda+\frac{1}{\delta_\circ})\int_Q u(x)^2 r(x)\,dx + \int_Q |Du(x)|^2 r(x)\,dx$$
$$+\int_Q f(x)^2 r(x)^\theta\,dx + \int_Q u(x)^2 dx + \int_{\partial Q}\phi(x)^2\,dS_x\Big].$$

Since $\Psi(y')=1$ for $|y'|<l_m$ and $0<y_n<h$, we get for $0<d\leqq h$

$$\sup_{0<\delta\leqq d}\int_{\partial Q_\delta} u^2 dS_x \leqq C_{12}\Big[(\lambda+\frac{1}{\delta_\circ^2})\int_Q u(x)^2 r(x)\,dx + \int_Q |Du(x)|^2 r(x)\,dx$$
$$+\int_Q f(x)^2 r(x)^\theta\,dx + \int_Q u(x)^2\,dx + \int_{\partial Q}\phi(x)^2\,dS_x\Big],$$

which combined with (10.24) gives

$$(10.25)\quad \sup_{0<\delta\leqq d}\int_{\partial Q_\delta} u(x)^2\,dS_x \leqq C_{13}\Big[\int_Q f(x)^2 r(x)^\theta\,dx + \int_Q u(x)^2 r(x)\,dx + \int_{\partial Q}\phi(x)^2\,dS_x\Big].$$

The result follows from (10.24) and (10.25).

As an immediate consequence of Theorem 10.1, we obtain

Theorem 10.2. *Let $\phi\in L^2(\partial Q)$. Then the Dirichlet problem (10.1), (10.2) admits the unique solution in $\widetilde{W}^{1,2}(Q)$ such that*

$$\int_Q |Du(x)|^2 r(x)\,dx + \int_Q u(x)^2 r(x)\,dx + \sup_{0<\delta<d}\int_{\partial Q_\delta} u(x)^2\,dS_x$$
$$\leqq C\Big[\int_Q f(x)^2 r(x)^\theta\,dx + \int_{\partial Q}\phi(x)^2\,dS_x\Big]$$

for some positive constant C and d.

10.3. C_{n-1}–estimate.

The aim of this section is to show that the norm $\|u\|_{C_{n-1}}$ of a solution of (10.1), (10.2) can be estimated by the L^2–norm of the boundary data ϕ and f. First we establish this estimate for the homogeneous equation (10.1). Also, we assume that $u \in W^{1,2}(Q)$; that is, ϕ is a trace of an element of the space $W^{1,2}(Q)$. By approximation we may assume that $\phi \in C^1(\partial Q)$. Obviously, the general case $\phi \in L^2(\partial Q)$ can be reduced to this situation. The crucial step is to show that

$$\int_Q u(x)^2 \, d\mu(x) \leqq \text{Const} \, \|\mu\| \int_{\partial Q} \phi(x)^2 \, dS_x \tag{10.26}$$

for all Carleson's measures μ. This estimate was obtained for analytic functions by Carleson [**CA1**] and subsequently extended to harmonic functions of several variable by Hörmander [**HO**].

We commence by introducing some terminology and notation. As in Section 10.1 we construct a covering $\{\Omega_k\}$, $k = 0, ..., p$, of the domain Q, where the sets have the following form

$$\Omega_k = \{(x', x_n);\ |x'| < l_l,\ F_k(x') < x_n < F_k(x') + \frac{l_k - |x'|}{a_1}\} \subset \Pi_k^{l_k, h},$$

$k = 1, ..., p$ and $\bar{\Omega}_\circ \subset Q$. The number $a_1 \geqq 2$, depending on $\|a^{ij}\|_\infty$, is to be determined later. We denote the corresponding partition of unity by $\{\zeta_k\}$, $k = 0, 1, ..., p$, that is; $\zeta_\circ \in C_\circ^\infty(\Omega_\circ)$, $\zeta_k \in C_\circ^\infty(\Omega_k')$, with

$$\Omega_k' = \{(x', x_n);\ |x'| < l_k,\ F_k(x') - \frac{l_k - |x'|}{a_1} < x_n < F_k(x') + \frac{l_k - |x'|}{a_1}\},$$

$k = 1, ..., p$ and $\sum_{k=0}^{p} \zeta_k(x) = 1$ on Q.

Let us now observe that if $u \in W^{1,2}(Q)$ is a solution of (10.1), (10.2), then $\zeta_\circ u$ is a solution of (10.1), with the right hand side $f_\circ \in W^{-1,2}(Q)$ given by

$$f_\circ(x) = D_j\big(a^{ij}(x) u D_j \zeta_\circ\big) - a^{ij}(x) D_i u D_j \zeta_p,$$

and

$$\|f_\circ\|_{W^{-1,2}} \leqq \text{Const} \, \|u\|_{W^{1,2}(\Omega_\circ)} \leqq \text{Const} \int_{\partial Q} \phi(x)^2 \, dS_x.$$

Let us now consider $u_k(x) = \zeta_k(x) u(x)$, $k = 1, ..., p$. The function u_k is a solution of the problem

$$\begin{aligned} -D_j\big(a^{ij}(x) D_i u_k\big) &= D_j\big(a^{ij}(x) u D_i \zeta_k\big) - a^{ij}(x) D_i u D_j \zeta_k \\ &= f_k(x) + g_k(x) \quad \text{in } Q \end{aligned}$$

and

$$u_k(x) = \zeta_k(x) \phi(x) \quad \text{on } \partial Q,$$

where f_k and g_k are in $W^{-1,2}(Q)$. For an arbitrary $\eta \in \overset{\circ}{W}{}^{1,2}(Q)$ we have

$$|\langle f_k, \eta\rangle| = |\int_{\Omega_k} u(x)a^{ij}(x)D_i\zeta_k D_j\eta\, dx| \leqq C\|u\|_2 |||\eta|||_2,$$

where a constant $C > 0$ depends on n, $\|a^{ij}\|_\infty$ and $\sup |D\zeta_k(x)|$, and

$$|\langle g_k, \eta\rangle| = \int_{\Omega_k} a^{ij}(x)D_i u D_j\zeta_k\eta\, dx|$$

$$\leqq C'\left(\int_{\Omega_k} \frac{\eta(x)^2}{r(x)}\, dx\right)^{\frac{1}{2}}\left(\int_{\Omega_k} |Du(x)|^2 r(x)\, dx\right)^{\frac{1}{2}},$$

where a constant $C' > 0$ is of the same nature as C. According to Theorem 2.2 we have

$$\|g_k\|_{W^{-1,2}}, \|f_k\|_{W^{-1,2}} \leqq \text{ Const } \int_{\partial Q} \phi(x)^2\, dS_x.$$

This shows that we can reduce the estimation of the norm $\|u\|_{C_{n-1}}$ of a solution of (10.1), (10.2) to that of the problem

$$-D_i\big(a^{ij}(x)D_j u\big) = 0 \quad \text{in } \Omega_k,$$

$$u(x) = \zeta(x)\phi(x) \quad \text{on } \partial\Omega_k \cap \partial Q \text{ and } u(x) = 0 \text{ on } \partial\Omega_k \cap Q.$$

We now fix Ω_k and for simplicity we write $\Omega = \Omega_k$. As in Section 9.3 we introduce the change of variables $y' = x'$, $y_n = x_n - F(x')$. The image of Ω is denoted again by Ω. The composition of u with this transformation is denoted again by u and the coefficients of the transformed equation are also denoted by a^{ij}. Since under this mapping, a Carleson measure is transformed into a Carleson measure with an equivalent norm, we may therefore restrict our attention to the Dirichlet problem in a cone $\Omega = \{(x', x_n);\ |x'| < l,\ 0 < x_n < \frac{l-|x'|}{a_1}\}$; that is,

$$-D_i\big(a^{ij}(x)D_j u\big) = 0 \quad \text{in } \Omega, \tag{10.27}$$

$$u(x) = 0 \text{ on } \partial\Omega \cap \{x_n > 0\} \text{ and } u(x) = \phi(x) \text{ on } \partial\Omega \cap \{x_n = 0\}, \tag{10.28}$$

with $\phi \in C^1(R_{n-1})$ and $\operatorname{supp}\phi \subset B_{n-1}(0,l)$. We extend u to R_n^+ by setting $u \equiv 0$ on $R_n^+ - \Omega$. It is well known that u is Hölder continuous on Ω and also u is continuous on $\{x;\ x_n 0 \geqq\}$. For an arbitrary $x' \in R_{n-1}$ we set,

$$\Gamma_1(x') = \{(y', y_n) \in R_n^+;\ |y' - x'| \leqq a_1 y_n\}$$

and we denote by $N(x')$, the maximal fumction of u defined by

$$N(x') = \max_{y \in \Gamma_1(x')} |u(y)|.$$

It is clear that $N \in C(R_{n-1})$ and $\operatorname{supp} N \subset \overline{B_{n-1}(0,l)}$. Let

$$\Gamma_2(x') = \{(y', y_n) \in R_n^+;\ |y' - x'| < a_2 y_n\},$$

where $a_2 = 2a_1$ and set

$$S(x') = \left(\int_{\Gamma_2(x')} y_n^{2-n}|Du(y', y_n)|^2 dy\right)^{\frac{1}{2}}.$$

The function $S(x')$ is called the area integral of u. We need a sharper form of Theorem 5.2, which involves the area integral.

Lemma 10.4. *Let $u \in W^{1,2}(\Omega)$ be a solution of (10.27). Then for all $x' \in B_{n-1}(0,l)$ and all $y = (y', y_n)$ and $z = (z', z_n)$ in $\Gamma_1(x')$ with $\frac{h}{2} \leqq y_n \leqq h$ and $\frac{h}{2} \leqq z_n \leqq h$, we have*

$$|u(y) - u(z)| \leqq C\frac{|y-z|^\alpha}{h^\alpha} S(x'), \tag{10.29}$$

where $C > 0$ and $\alpha > 0$ are constants depending on n, γ and $\|a^{ij}\|_\infty$.

PROOF. It is obvious that it suffices to establish this inequality for $h = 1$. Let

$$\Gamma_1'(x') = \Gamma_1(x') \cap \Omega \cap \{(y', y_n);\ \frac{1}{2} \leqq y_n \leqq 1\}$$

and

$$\Gamma_2'(x') = \Gamma_2(x') \cap \Omega \cap \{(y', y_n);\ \frac{1}{4} < y_n < \frac{5}{4}\}$$

and suppose that y and z are in $\Gamma_1'(x')$ and that outside Ω $u \equiv 0$. The inequality (10.29) is invariant under translation of u. Therefore we prove it for $\tilde{u}(y) = u(y) - c_\circ$, with

$$c_\circ = \frac{1}{|\Gamma_2(x')|} \int_{\Gamma_2'(x')} u(y)\, dy.$$

By Theorem 5.2 (see Chapter 5), for any solution of (10.27) with $\tilde{u} =$ const on $\partial\Omega \cap \Gamma_1'(x')$, we have

$$\|\tilde{u}\|_{C^\alpha(\overline{\Gamma_1'(x')})} \leqq C_\circ \|\tilde{u}\|_{L^2(\Gamma_2'(x'))},$$

where the constant $C_\circ > 0$ depends on n, γ and $\|a^{ij}\|_\infty$. Consequently, by Poincaré's inequality, we get

$$\begin{aligned}\frac{|\tilde{u}(y) - \tilde{u}(z)|}{|y-z|^\alpha} &\leqq \|\tilde{u}\|_{C^\alpha(\Gamma_1'(x'))} \leqq C_\circ \|\tilde{u}\|_{L^2(\Gamma_2'(x'))} \leqq C_\circ \||D\tilde{u}|\|_{L^2(\Gamma_2'(x'))} \\ &\leqq \left(\frac{5}{4}\right)^{\frac{n-2}{2}} \left(\int_{\Gamma_2(x')} y_n^{2-n} |D\tilde{u}(y)|^2\, dy\right)^{\frac{1}{2}}\end{aligned}$$

for some constant $C_\circ' > 0$ and the results follow easily.

To proceed further we set

$$E(\lambda) = \{x' \in R_{n-1};\ N(x') > \lambda\}.$$

It is obvious that $\overline{E(\lambda)} \subset B_{n-1}(0,l)$. Let $\kappa > 0$, by $E'(\kappa\lambda)$ we denote the set of points $x' \in E(17\lambda)$ having the following property:

for each ball $B \subset R_{n-1}$, containing x', we have

$$|\{y' \in B;\ S(y') > \kappa\lambda\}| \leqq 2^{-n}|B|. \tag{10.30}$$

Lemma 10.5. *If $x' \in E'(\kappa\lambda)$, then*

$$S(x') \leqq \kappa\lambda.$$

PROOF. In the contrary case $S(x') > \lambda\kappa$ and there exists a $\delta > 0$ such that

$$S_\delta(x') = \int_{\Gamma_2(x')} (y_n + \delta)^{2-n} |Du(y)|^2 \, dy > (\kappa\lambda)^2.$$

The absolute continuity of the Lebesgue integral implies the existence of a ball $B_{n-1}(x', r)$ such that

$$S(y') \geqq S_\delta(y') \geqq \kappa\lambda$$

for all $y' \in B_{n-1}(x', r)$. This means that $x' \in E'(\kappa\lambda)$ and we have arrived at the contradiction.

Lemma 10.6. *Let $B = B_{n-1}(0,1)$ and $V = \{x = (x', x_n); \, |x'| < 1 - a_1 x_n, \, 0 < x_n < \frac{1}{a_1}\}$. Suppose that $\bar{B} \subset \overline{E(\lambda)}$ and $|B \cap E'(\lambda\kappa)| > 0$. Then there exists a constant $C_1 = C_1(n)$, depending on n, such that*

$$\int_Z y_n |Du(y)|^2 \, dy \leqq \frac{C_1(n)}{a_1^{n-1}} \kappa^2 \lambda^2,$$

where $Z = V \cap \bigcup_{x' \in \cap E'(\lambda\kappa)} \Gamma_2(x')$.

PROOF. It follows from Lemma 10.5 that

$$(10.32) \quad C(n)\kappa^2\lambda^2 > \int_{x' \in B, \, S(x) \leqq \lambda\kappa} S(x')^2 \, dx'$$

$$= \int_{x' \in B, \, S(x') \leqq \kappa\lambda} \int_{\Gamma_2(x')} y_n^{2-n} |Du(y)|^2 \, dy dx'$$

$$\geqq \int_Z y_n^{2-n} |Du(y)|^2 |\{x' \in B, \, S(x') \leqq \lambda\kappa, \, |x' - y'| < a_2 y_n\}| \, dy.$$

For each $y \in Z$ there exists $\xi' \in B \cap E'(\kappa\lambda)$ such that $|y' - x'| < a_2 y_n$. Hence applying (10.32) with $B(y', a_2 y_n)$, we get

$$|\{x' \in B(y', a_2 y_n); \, S(x') \leqq \kappa\lambda\}| \geqq (1 - 2^{-n}) |B(y', a_2 y_n)|.$$

Since $a_2 = 2a_1$, we get

$$|\{x'; \, a_1 y_n < |y' - x'| < a_2 y_n\}| = c_{n-1}(a_2^{n-1} - a_1^{n-1}) y_n^{n-1}$$
$$= (1 - 2^{-(n-1)}) |B(y', a_2 y_n)|.$$

Since $y = (y', y_n) \in V$, we have $B(y', a_1 y_n) \subset B$ and hence

$$|\{x' \in B; \, x' \in B(y', a_2 y_n), \, S(x') \leqq \kappa\lambda\}| \geqq |\{x' \in B(y', a_2 y_n); \, S(x') \leqq \kappa\lambda\}|$$
$$- |\{x'; \, a_1 y_n < |x' - y'| < a_2 y_n\}|$$
$$\geqq \left((1 - 2^{-n}) - (1 - 2^{-(n-1)})\right) |B(y', a_2 y_n)|$$
$$= 2^{-n} c_{n-1} a_2^{n-1} y_n^{n-1} = \frac{c_{n-1}}{2} (a_1 y_n)^{n-1},$$

where c_{n-1} denotes the volume of a unit ball in R_{n-1}. When combined with (10.31), this gives (10.32).

Let B and V be defined as in Lemma 10.6 and suppose that $\bar{B} \subset \overline{E(\lambda)}$, with $B \cap \partial E(\lambda) \neq \emptyset$. Let us set $E_\circ = E'(\kappa\lambda) \cap B(0, 1-2\eta)$ and suppose that for some $0 < \eta < \frac{1}{2}$ and $C > 0$ we have

$$(10.33) \qquad |E_\circ| \geqq C|B|.$$

Finally, let us set $W_1 = V \cap \bigcup_{x' \in E_\circ} \Gamma_1(x')$. Since $B \cap \partial E(\lambda) \neq \emptyset$, there exists a half-line H emanating from this intersection and passing through the vertex of V, such that

$$u(y) \leqq \lambda \quad \text{for } y \in H.$$

Consequently, applying Lemma 10.4 finitely many times, we arrive at the estimate

$$|u(x)| \leqq \lambda + C'(\eta, \gamma, \|a^{ij}\|_\infty) a_1^\alpha \kappa\lambda,$$

for all $y' \in E_\circ$ and all $x \in \Gamma_1(y')$ with $x_n > \frac{\eta}{a_1}$. Imposing the condition

$$(10.36) \qquad C'(n, \gamma, \|a^{ij}\|_\infty) a_1^\alpha \kappa < 2,$$

we get

$$(10.35) \qquad \sup\{|u(x)|; x \in \bigcup_{y' \in E_\circ} \Gamma_1(x') \text{ and } x_n > \frac{\eta}{a_1}\} < 3\lambda.$$

To proceed further, we set

$$\Gamma_1(x) = \{y = (y', y_n);\ y_n \geqq x_n,\ |y' - x'| \leqq a_1(y_n - x_n)\},$$

$$N(x) = \sup_{y \in \Gamma_1(x)} |u(y)|$$

and

$$W = \{x \in W_1;\ N(x) < 9\lambda\}.$$

It follows from the inequality (10.35) that

$$\{x \in W,\ x_n \geqq \frac{\eta}{a_1}\} \subset W \subset W_1.$$

Each line parallel to the x_n–axis intersects ∂W at most two points, one of them with a larger x_n–coordinate belongs to the lateral surface of the cone V. We denote this set by ∂W^+, that is, $\partial W^+ = \partial W \cap \partial V$. Obviously $\partial W \cap B = \emptyset$. The x_n–coordinate of a point $x \in \partial W^+$ satisfies the inequality $x_n \geqq \frac{\eta}{a_1}$ and by (10.36) we have

$$(10.36) \qquad |u(x) \leqq 3\lambda \quad \text{for all } x \in \partial W^+.$$

We set $\partial W^- = \partial W - \partial W^+$. Let $P(W)$ be the projection of W on the plane $\{x_n = 0\}$. Then $P(W) \subset B(0, 1-\eta)$. Writing

$$t(x') = \inf\{x_n;\ (x', x_n) \in W\},$$

we have

$$\partial W^- = \{(x', x_n);\ x' \in P(W),\ x_n = t(x')\}$$

and

$$(10.37) \qquad |t(x') - t(y')| \leqq \frac{1}{a_1}|x' - y'|$$

for all x' and y' in $P(W)$.

Lemma 10.7. *Let $\Sigma = \{x \in \partial W^-;\ |u(x)| > 5\lambda\}$ and denote by $P(\Sigma)$ the projection of Σ on $\{x_n = 0\}$. Suppose that the condition (10.33) holds. Then for κ sufficiently small there exists a constant $c = c(n) > 0$ such that*

$$|P(\Sigma)| \geqq c|B|. \tag{10.38}$$

PROOF. Let $x' \in E_\circ$. Since $N(x') = N((x',0)) = \sup_{y\in\Gamma_1(x'),\, y_n < \frac{\eta}{a_1}} |u(y)| > 17\lambda$, according to (10.36) we have

$$x' \in P(W) \cap B_\circ \quad \text{and} \quad N((x',t(x'))) = 9\lambda.$$

Consequently, there exists $y = (y', y_n)$ with $|x' - y'| = a_1(y_n - t(x'))$ and $|u(y)| = 9\lambda$. Therefore $y' \in P(W)$, $y_n = t(y') < \frac{\eta}{a_1}$, $|y' - x'| \leqq a_1 y_n \leqq \eta$ and hence $B_{n-1}(x'; |y' - x'|) \subset P(W)$. By (10.37), if $x' \in B(x', |x' - y'|) \cap B(y', \frac{a_1}{2} t(y'))$, then $t(x') \in (\frac{t(y')}{2}, t(y'))$. Hence Lemma 10.4 yields

$$\begin{aligned} |u(y) - u(z)| &\leqq |u(y',t(y')) - u(z',t(y'))| + |u(z',t(y')) - u(z',t(z'))| \\ &\leqq C\left[\frac{|y'-z'|^\alpha}{t(y')^\alpha} S(x') + \frac{|t(y') - t(z')|}{t(y')^\alpha} S(\xi')\right] \leqq C\left[(\frac{a_1}{2})^\alpha + (\frac{1}{2})^\alpha\right]\kappa\lambda, \end{aligned}$$

where $\xi' \in E_\circ$ with $z = (z', t(z'))$ and $(z', t(y'))$ in $\Gamma_1(\xi')$. As a consequence of (10.34) we have

$$\kappa C(\frac{a_1}{2}) < 2.$$

Therefore $|u(z)| > 5\lambda$ and

$$B(x', |y' - x'|) \cap B(y', \frac{a_1}{2} t(y')) \subset P(\Sigma).$$

We now observe that for each $x' \in E_\circ$ there is a $\rho = |x' - y'| > 0$ such that

$$\frac{1}{|B(x',\rho)|} \int_{B(x',\rho)} \chi_{P(Z)}(z')\, dz' \geqq \sigma,$$

where

$$\sigma = \sigma(n) = \frac{|B_{n-1}(0,1) \cap B(y', \frac{1}{2})|}{|B_{n-1}(0,1)|}$$

with $|y'| = 1$. In other words, if $x' \in E_\circ$ then $M\chi_{P(\Sigma)} \geqq \sigma$. Consequently, by (10.33) we have

$$c|B| \leqq |E_\circ| \leqq |\{x' \in B;\ M\chi_{P(\Sigma)}(x') \geqq \sigma\}| \leqq \frac{1}{\sigma} C_\circ(n)|P(\Sigma)|,$$

and this completes the proof.

The following lemma is a consequence of Whitney's theorem (for details see Lemma 4 in [BU]).

Lemma 10.8. *Let G be an open bounded and nonempty subset of R_n and let $F = R_n - G$. Let $\alpha > 1$ and suppose that E is a measurable subset of G satisfying $|G| \leqq \alpha |E|$. Then there is a ball $B \subset G$ with $\partial B \cap F \neq \varnothing$, and such that*

$$|B| \leqq c\alpha |E \cap B|,$$

where a constant $c > 0$ depends only on n.

We are now in a convenient position to establish the key relation between sets $E(\lambda)$ and $E'(\kappa\lambda)$ that is needed to prove the inequality (10.26).

Lemma 10.9. *There exist positive constants a_1 and κ depending on n, γ and $\|a^{ij}\|_\infty$ such that*

$$|E(\lambda)| \leqq 2 \cdot 17^2 |E'(\kappa\lambda)| \tag{10.40}$$

for all $\lambda > 0$.

PROOF. Suppose that (10.40) is not true. Then there exists a $\lambda > 0$ such that

$$|E(\lambda)| < 2 \cdot 17^2 |E'(\kappa\lambda)|$$

for all $\kappa > 0$. It follows from Lemma 10.8 that there exists a constant $C = C(n)$ and a ball $B \subset R_{n-1}$ such that $\bar{B} \subset \overline{E(\lambda)}$, $B \cap \partial E(\lambda) \neq \varnothing$ and

$$|B| \leqq 2 \cdot 17^2 C |B \cap E'(\kappa\lambda)|.$$

We may assume that $B = B_r = B_{n-1}(0, r)$. We now choose $\eta > 0$ so that

$$|B_r - B_\circ| \leqq \frac{1}{24^2 C} |B_r|,$$

where $B_\circ = B_{n-1}(0, (1 - 2\eta)r)$. Writing $E_\circ = B_\circ \cap E'(\kappa\lambda)$, we get

$$|E_\circ| \geqq \frac{1}{34^2 C} |B_r|. \tag{10.41}$$

Without loss of generality, we may assume that $r = 1$. Consequently, we obtain a situation in which both Lemmas 10.6 and 10.7 are valid. This can be achieved by introducing new variables $y = \frac{x}{r}$. We denote the coefficients of the transformed equation again by a^{ij}. Then their modulus of continuity $\tilde{\omega}_r$ is given by $\tilde{\omega}_r(t) = \omega(rt)$. For $r < l$, we have

$$\tilde{\omega}_r\left(\frac{1}{a_1}\right) = \omega\left(\frac{r}{a_1}\right) \leqq \omega\left(\frac{l}{a_1}\right)$$

and

$$\int_0^{\frac{1}{a_1}} \frac{\tilde{\omega}_r(t)}{t}\, dt = \int_0^{\frac{r}{a_1}} \frac{\omega(t)}{t}\, dt \leqq \int_0^{\frac{l}{a_1}} \frac{\omega(t)}{t}\, dt.$$

From now on, we denote the modulus of continuity of coefficients a^{ij} of the transformed equation again by ω. It is clear that ω satisfies

$$\max\{\frac{1}{\omega(a_1)}, \int_0^{\frac{1}{a_1}} \frac{\omega(t)}{t}\, dt\} \leqq \epsilon(a_1), \tag{10.42}$$

with $\epsilon(a_1) \to 0$ as $a_1 \to \infty$. To complete the proof, we define the function $u^{3\lambda}$ by

$$u^{(3\lambda)}(x) = \begin{cases} u(x) - 3\lambda & \text{for } u(x) > 3\lambda, \\ 0 & \text{for } |u(x)| < 3\lambda, \\ u(x) + 3\lambda & \text{for } u(x) < -3\lambda. \end{cases}$$

Let us denote by $\Psi(x', x_n)$ a function with properties $\Psi(x', x_n) = 1$ for $x \in W$, with $x_n \geqq 2t(x')$, $\Psi(x', x_n) = \frac{x_n}{t(x')} - 1$ for $x \in W$, with $t(x') < x_n < 2t(x')$ and $\Psi(x', x_n) = 0$ for $x \in V - W$. It is obvious that $|u^{(3\lambda)}(x)| < 6\lambda$ for $x \in W$ and $u^{(3\lambda)}(x) = 0$ for $x \in \partial W^+$, moreover $\eta(x) = u^{(3\lambda)}(x) x_n \Psi(x)$ belongs to $\overset{\circ}{W}{}^{1,2}(V)$ with supp $\eta \subset \bar{W} \cap V$. Since $t(x')$ is a Lipschitz function (see (10.37) we easily verify that

$$\frac{\partial \Psi(x)}{\partial x_n} = \frac{1}{t(x')}, \quad \sum_{k=1}^{n-1} |D_k \Psi(x)|^2 \leqq \frac{x_n^2}{a_1^2 t(x')^4} \tag{10.43}$$

for all $x = (x', x_n) \in W$ with $x_n < 2t(x')$. Hence

$$|D\Psi(x)| \leqq \frac{\sqrt{2}}{t(x')} \tag{10.44}$$

and $|D\Psi(x)| = 0$ for all remaining points $x \in V$. Taking η as a test function in (10.27), we get

(10.45)
$$\begin{aligned}
0 &= \int_W [a^{ij} D_i u D_j u^{(3\lambda)} x_n \Psi + x_n u^{(3\lambda)} a^{ij} D_i u D_j \Psi \\
&+ u^{(3\lambda)} \Psi a^{ij} D_i u D_j x_n]\, dx = \int_W x_n \Psi a^{ij} D_i u D_j u^{(3\lambda)}\, dx \\
&+ \int_{P(W)} dx_n \int_{t(x')}^{2t(x')} x_n u^{(3\lambda)} a^{ij} D_i \Psi D_j u\, dx' + \int_W u^{(3\lambda)} \Psi (a^{ij} - a_\circ^{ij}) D_i x_n D_j u^{(3\lambda)}\, dx \\
&+ \frac{1}{2} \int_W \Psi \sum_{k=1}^{n-1} a_\circ^{kn} D_k \big(k u^{(3\lambda)}\big)^2\, dx + \frac{1}{2} \int_W a_\circ^{nn} D_k \big(u^{(3\lambda)}\big)^2 \Psi\, dx \\
&= I_1 + I_2 + I_3 + I_4 + I_5.
\end{aligned}$$

We now proceed to estimate the integrals I_i. Since $W \subset W_1 \subset Z$, it follows from Lemma 10.6 that

$$0 < I_1 \leqq \gamma_1 \int_Z x_n |Du(x)|^2\, dx \leqq \frac{C_1(n)}{a_1^{n-1}} \gamma_1 \kappa^2 \lambda^2,$$

where $\gamma_1 > 0$ depends on $\|a^{ij}\|_\infty$. Setting $W_2 = V \cap \bigcup_{x' \in B_\circ \cap E'(\lambda\kappa)} \Gamma_2(x')$, we have $W_2 \subset Z$ and by Lemma 10.6 and (10.44), we get

$$
\begin{aligned}
|I_2| &\leqq C(\gamma_1) \int_{P(W)} dx_n \int_{t(x')}^{2t(x')} \frac{x_n \lambda}{t(x')} |Du(x)|\, dx' \\
&\leqq C(\gamma_1)\lambda \Big(\int_{W_2} x_n |Du(x)|^2\, dx \Big)^{\frac{1}{2}} \Big(\int_{P(W)} dx_n \int_{t(x')}^{2t(x')} \frac{x_n}{t(x')^2}\, dx' \Big)^{\frac{1}{2}} \\
&\leqq \sqrt{\frac{3}{2}} |B| C(\gamma_1)\lambda \frac{C_1(n)^{\frac{1}{2}} \kappa\lambda}{a_1^{\frac{n-1}{2}}} = \frac{C_2(\gamma_1, n)\kappa\lambda^2}{a_1^{\frac{n-1}{2}}}
\end{aligned}
$$

By virtue of (10.10), (10.42) and Lemma 9.6, we have

$$
|I_3| \leqq 6\lambda \Big(\int_{W_2} x_n |Du(x)|^2\, dx \Big)^{\frac{1}{2}} \Big(\int_V \frac{\omega(x_n)^2}{x_n}\, dx \Big)^{\frac{1}{2}} \leqq \frac{C_3(n)\kappa\lambda^2 \epsilon(a_1)}{a_1^{\frac{n-1}{2}}}.
$$

It follows from (10.10), (10.42) and (10.44) that

$$
\begin{aligned}
|I_4| &\leqq \frac{1}{2} \Big| \int_W u^{(3\lambda)}(x)^2 \sum_{k=1}^{n-1} [a_\circ^{nk}(x) D_k \Psi(x) + \Psi(x) D_k a_\circ^{nk}(x)]\, dx \Big| \\
&\leqq C_4(n, \gamma_1)\lambda^2 \left[\frac{1}{a_1} + \int_{P(W)} dx' \int_0^{\frac{1}{a_1}} \frac{\omega(t)}{t}\, dt \right] \leqq C_4(n, \gamma_1) \Big(\frac{1}{a_1} + \epsilon(a_1) \Big) \lambda^2.
\end{aligned}
$$

To estimate I_5, we use the fact that $u^{(3\lambda)}(x) = 0$ on ∂W^+, while $\Psi(x) = 0$ on ∂W^-, obtaining

$$
\begin{aligned}
I_5 &= \frac{1}{2} \int_W \Psi(x) a^{nn}(x', 0) D_n u^{(3\lambda)}(x)^2\, dx = -\frac{1}{2} \int_W a^{nn}(x', 0) u^{(3\lambda)}(x)^2 D_n \Psi\, dx \\
&- \frac{1}{2} \int_{P(W)} a^{nn}(x', 0)\, dx_n \int_{t(x')}^{2t(x')} u^{(3\lambda)}(x)^2 D_n\, dx.
\end{aligned}
$$

Hence by (10.43), we get

$$
-I_5 \geqq \frac{\gamma}{2} \int_{P(\Sigma)} \left[\int_{t(x')}^{2t(x')} u^{(3\lambda)}(x)^2\, dx_n \right] t(x')^{-1}\, dx'.
$$

By Lemma 10.4 and the fact that $|u^{(3\lambda)}(x)| > 5\lambda - 3\lambda = 2\lambda$ for $x = (x', t(x')) \in \Sigma$, we obtain

$$
\begin{aligned}
|u^{(3\lambda)}(x)| &\geqq |u^{(3\lambda)}(x', t(x'))| - |u(x', x_n) - u(x', t(x'))| \\
&\geqq 2\lambda - C \Big(\frac{|x_n - t(x')|}{t(x')} \Big)^\alpha S(\xi)
\end{aligned}
$$

for $\xi \in E_o$ and $(x', t(x')) \in \Gamma_1(\xi)$. Hence, by Lemma 10.5, for κ sufficiently small, (say $\kappa C < 1$), we have $|u^{(3\lambda)}(x)| > \lambda$ for $t(x') < x_n < 2t(x')$. Consequently

$$-I_5 \geqq \frac{\gamma}{2}\lambda^2 \int_{P(\Sigma)} dx' \int_{t(x')}^{2t(x')} dx_n = \frac{\gamma\lambda^2}{2}|P(\Sigma)|$$

and by Lemma 10.7

$$-I_5 \geqq \frac{C(n)\gamma}{2}|B|\lambda^2 = C_5(n,\gamma)\lambda^2.$$

Inserting the estimate for I_5 into (10.45) we obtain

$$C_5(n,\gamma)\lambda^2 \leqq -I_5 = I_1 + I_2 + I_3 + I_4 \leqq \frac{C_1(n)\gamma_1}{a_1^{n-1}}\kappa\lambda^2$$
$$+ \frac{C_2(n,\gamma_1)}{a_1^{\frac{n-1}{2}}}\kappa\lambda^2 + \frac{C_3(n)\epsilon(a_1)}{a^{\frac{n-1}{2}}}\kappa\lambda^2 + C_4(n,\gamma_1)\left(\frac{1}{a_1} + \epsilon(a_1)\right)\lambda^2.$$

We now choose $a_1 = a_1(n,\gamma,\gamma_1)$ sufficiently large, that is, $\epsilon(a_1)$ is small, and then for a fixed a_1 we choose κ sufficiently small to obtain a contradiction with the last inequality and this completes the proof of Lemma 10.10.

We are in a position to establish the basic estimate (10.26).

Theorem 10.3. *Let $u \in W^{1,2}(Q)$ be a solution of the problem* (10.1), (10.2). *Then for every Carleson measure we have*

$$\int_Q u(x)^2\, d\mu(x) \leqq C\|\mu\| \int_{\partial Q} \phi(x)^2\, dS_x,$$

where $C > 0$ is a constant independent of u, ϕ and μ.

PROOF. According to the remarks made at the beginning of this section, it suffices to consider a solution $u \in W^{1,2}(\Omega)$ of the problem (10.27), (10.28). We have

$$\int_{R_{n-1}} S(x')^2\, dx' = \int_{R_{n-1}} \left(\int_{\Gamma_2(x')} y_n^{2-n}|Du(y', y_n)|^2\, dy' dy_n\right) dx'$$
$$= C(n) \int_{R_{n-1}} y_n^{n-2}(a_2 y_n)^{n-1}|Du(y)|^2\, dy$$
$$= C(n)a_2^{n-2} \int_{R_n} y_n|Du(y)|^2\, dy \leqq C(n,a_1) \int_{\partial Q} \phi(x')^2\, dS_x.$$

Since $N(x') > \lambda$ is equivalent to $\Gamma_1(x') \cap \{y \in R_n^+;\ |u(y)| > \lambda\} \neq \varnothing$, we get for any Carleson measure μ, corresponding to $h(r) = r^{n-1}$, that

$$\mu(\{x \in R_n^+;\ |u(x)| > \lambda\}) \leqq C(n,a_1)\|\mu\||\{x' \in R_{n-1};\ N(x') > \lambda\}|.$$

Consequently

$$\begin{aligned}\int_{R_n^+} u(x)^2\, d\mu(x) &= \int_0^\infty \mu(\{x \in R_n^+;\ |u(x)| > \lambda\})|\, d\lambda^2 \\ &\leqq C(n,a_1)\|\mu\| \int_0^\infty |\{x' \in R_{n-1};\ N(x') > \lambda\}|\, d\lambda^2 \\ &= C(n,a_1)\|\mu\| \int_{R_{n-1}} N(x')^2\, dx'.\end{aligned}$$

To complete the proof, we must show that

$$\int_{R_{n-1}} N(x')^2\, dx' \leqq \text{Const} \int_{R_{n-1}} S(x')^2\, dx'. \tag{10.46}$$

Let $\lambda > 0$. Then for each $x' \in E(17\lambda)$, either $x' \in E'(\kappa\lambda)$ or $M\chi(x') > 2^{-n}$, where $M\chi$ is the Hardy– Littlewood maximal function of the characteristic function of the set $\{y' \in R_{n-1};\ S(y') > 2^{-n}\}$. Hence, by the well–known property of the Hardy–Littlewood maximal function, we have

$$\begin{aligned}|E(17\lambda)| &\leqq |E'(\kappa\lambda)| + |\{x' \in B_{n-1}(0,l);\ M\chi(x') > 2^{-n}\}| \\ &\leqq |E'(\kappa\lambda)| + 2^n C_\circ \|\chi\|_{1,R_{n-1}} \\ &= |E'(\kappa\lambda)| + 2^n C_\circ |\{x' \in R_{n-1};\ S(x') > \kappa\lambda\}|,\end{aligned}$$

where $C_\circ > 0$ is an absolute constant. Consequently by Lemma 10.9, we obtain

$$\begin{aligned}&\int_{R_{n-1}} N(x')^2\, dx' = 2\int_0^\infty \lambda |E(\lambda)|\, d\lambda = 2\cdot 17^2 \int_0^\infty t|E(17t)|\, dt \\ &\leqq 2\cdot 17^2 \int_0^\infty t|E'(\kappa t)|\, dt + 2\cdot 17^2 2^n C_\circ \int_0^\infty t|\{x' \in R_{n-1};\ S(x') > \kappa t\}|\, dt \\ &\leqq \frac{1}{2}\int_0^\infty |E(t)|\, dt^2 + \frac{2^n C_\circ 17^2}{\kappa^2} \int_{R_{n-1}} S(x')^2\, dx' \\ &= \frac{1}{2}\int_{R_{n-1}} N(x')^2\, dx' + C(n,\gamma,\gamma_1)\int_{R_{n-1}} S(x')^2\, dx',\end{aligned}$$

which implies (10.46).

Theorem 10.3 yields the following estimate of u in norm $\|\cdot\|_{C_{n-1}}$.

Theorem 10.4. *Let $\phi \in L^2(\partial Q)$ and $f \in W^{-1,2}(Q)$. Then the problem (10.1),(10,2) has a unique solution in $\widetilde{W}^{1,2}(Q)$ satisfying the estimate*

$$\|u\|_{C_{n-1}} + \int_Q |Du(x)|^2 r(x)\, dx \leqq C\Big(\int_{\partial Q} \phi(x)^2\, dS_x + \|f\|_{W^{-1,2}}\Big) \tag{10.47}$$

for some constant $C > 0$.

PROOF. First we solve the Dirichlet problem with $\phi \equiv 0$ on ∂Q. Then the solution u belongs to $W^{1,2}(Q)$ and by Theorem 9.2 $W^{1,2}(Q) \subset C_{n-1}(\bar{Q})$. Therefore (10.47) holds in this case. Consequently it suffices to establish (10.47) for the homogeneous equation (10.1) ($f \equiv 0$ on Q) with $\phi \in L^2(\partial Q)$. By approximation this can be reduced to the case $\phi \in C^1(\partial Q)$ and the result easily follows from Theorem 10.3.

10.4. Remarks on the absolute continuity of harmonic measures.

In Chapter 5 (Theorem 5.7), under regularity assumptions on coefficients and ∂Q, we proved that the harmonic measure W_L^x associated with the operator L and the Lebesgue surface measure are mutually absolutely continuous. This result, specialized to L given by (10.1), says that if ∂Q is of class C^2 and the coefficients satisfy Dini condition on $\bar{Q}$, then the harmonic measure W_L^x and the Lebesgue surface measure are mutually absolutely continuous. The essential ingredient of the proof was the energy estimate, which by Theorem 10.2 continues to hold if a^{ij} are Dini on ∂Q and ∂Q is of class $C^{1,\mathrm{Dini}}$. Therefore we can state the following

Theorem 10.5. *Let (A_1) and (A_2') hold and suppose that ∂Q is of class $C^{1,\mathrm{Dini}}$. Then the harmonic measure W_L^x associated with the operator L, given by (10.1), and the Lebesgue surface measure are mutually absolutely continuous.*

We emphasize here, that we do not require continuity of a^{ij} on $\bar{Q}$. Theorem 10.5 is closely related to Theorem 1 in [**F,J,K**], where a similar result was obtained under the additional assumption that $a^{ij} \in C(\bar{Q})$, $i,j = 1,...,n$. However, the assumption on ∂Q and the Dini condition on a^{ij} on ∂Q are slightly weaker; namely, the authors of [**F,J,K**] assume that ∂Q is of class C^1 and the coefficients a^{ij} are Dini on ∂Q in a certain prescribed non tangential direction.

We point out here that the recent results of Dahlberg [**DA**] and R.Fefferman [**FE**] also show that the assumption of continuity of a^{ij} on $\bar{Q}$ is not needed for the absolute continuity of harmonic measure. Their method is based on the comparison of harmonic measures corresponding to two different elliptic operators.

To describe these results, we recall some definitions. Let $Q = B(0,1)$. If $\Delta \subset \partial B$, is a surface ball of radius r, centered at $x \in \partial B$, we set $S(\Delta) = B(x,r) \cap B$. In his approach, Dahlberg [**DA**] used the following version of a Carleson measure: A nonnegative measure μ on B is said to be a Carleson measure if $\mu(S(\Delta)) \leqq \mathrm{Const}\, \sigma(\Delta)$ for all surface balls $\Delta \subset \partial B$, where σ denotes the Lebesgue surface measure on ∂B. We say that μ is a Carleson measure with vanishing trace provided there exists a function $h : (0,1) \to (0,\infty)$, such that $\lim_{r\to 0} h(r) = 0$ and $\mu(S(\Delta)) \leqq h(r)\sigma(\Delta)$ for all surface balls Δ of radius r, and $0 < r < 1$.

The following result is due to Dahlberg [**DA**].

Theorem 10.6. *Suppose that L and $\bar{L}$ are two elliptic operators satisfying (A_1) with coefficients matrices $A(x) = \{a^{ij}(x)\}$ and $\bar{A}(x) = \{\bar{a}^{ij}(x)\}$, respectively. For $y \in B$ we set*

$$\eta(y) = \sup_{x \in B(y, \frac{r(y)}{2})} \|A(x) - \bar{A}(x)\|. \tag{10. 48}$$

Assume that $\frac{\eta(y)}{r(y)}\, dy$ is a Carleson measure with vanishing trace. If the harmonic measure W_L^x is absolutely continuous with respect to the Lebesgue surface measure, then the same is true for the harmonic measure $W_{\bar{L}}^x$.

The condition (10.48) means that the coefficients a^{ij} are uniformly close to $\bar{a}^{ij}$ as we approach the boundary ∂B. R.Fefferman **[FE]** relaxed that condition by assuming only that there exists a constant $C > 0$ such that

$$\int_0^1 \frac{\eta((1-t)x)}{t}\, dt \leqq C$$

for all $x \in \partial B$. It is clear that results of Dahlberg and Fefferman also show that the continuity of a^{ij} on $\bar{B}$ is not needed for absolute continuity of the harmonic measure. The methods employed in papers **[FJK]**, **[DA]** and **[FE]** are very different and go far beyond the scope of this book.

BIBLIOGRAPHY

[AD] R.A. Adams, *Sobolev spaces*, Academic Press, New York - San Francisco - London, 1975.

[BG] D.L.Burkholder & R.F.Gundy, *Distribution function inequalities*, Studia Math. **46** (1972), 527–544.

[BM] G. Bottaro & M.E. Marina, *Probleme di Dirichlet per equazioni ellittiche di tipo variazionale su insiemi non limitati*, Boll. Un. Math. Ital. **8** (1973), 46–56.

[BI1] A.V. Bitsadze, *On the theory of nonlocal boundary value problems*, Soviet Mat. Dokl. **30(1)** (1984), 8–10.

[BI2] A.V. Bitsadze, *On a class of conditionally solvable nonlocal boundary value problems for harmonic functions*, Soviet Mat. Dokl. **31(1)** (1985), 91–94.

[BS] A.V. Bitsadze & A.A, Smarskiĭ, *On some simple generalizations of linear elliptic boundary problems*, Soviet Math. Dokl. **10(2)** (1969), 398–400.

[CFK] L.Caffarelli, E.Fabes & C.Kenig, *Completely singular elliptic-harmonic measures*, Indiana Univ. Math. J. **30** (1981), 917–924.

[CA] L.Carleson, *Selected problems on exceptional sets*, Van Nostrand, Princeton, N.J., 1967.

[CA1] L.Carleson, *Interpolations by bounded analytic functions and the Corona problem*, Ann. of Maths **76(3)** (1962), 123–131.

[CT1] J.Chabrowski & H.B. Thompson, *On the boundary values of the solutions of linear elliptic equations*, Bull. Austral. Math. Soc. **27** (1983), 1–30.

[CT2] J.Chabrowski & B.Thompson, *On traces of solutions of a semilinear partial differential equation of elliptic type*, Ann. Pol. Math. **42** (1983), 45–71.

[CH1] J.Chabrowski, *Note on the Dirichlet problem with L^2-boundary data*, Manuscripta Math. **40** (1982), 91–108.

[CH2] J.Chabrowski, *Dirichlet problem for a linear elliptic equation in unbounded domains with L^2-boundary data*, Rend. Sem. Mat. Univ. Padova. **71** (1984), 287–328.

[CH3] J.Chabrowski, *On the Dirichlet problem for a linear elliptic equation with unbounded coefficients*, Boll. Un. Math. Ital. **5–B** (1986), 71–91.

[CH4] J.Chabrowski, *On the Dirichlet problem for a degenerate elliptic equation*, Com. Math. Univ. Carolinae **28** (1987), 141–155.

[CH5] J.Chabrowski, *On solvability of boundary value problem for elliptic equations with Bitsadze–Samarskii condition*, Internat. J. Math. & Math. Sci. **11** (1988), 101–114.

[CH6] J.Chabrowski, *On the Dirichlet problem with L^1-boundary data*, Funkcialaj Ekv. **28** (1985), 327–339.

[CH7] J.Chabrowski, *On the Dirichlet problem for degenerate elliptic equations*, Publ. Research Instit. Math. Sci. Kyoto Univ. **23** (1987), 1–16.

[CL] J.Chabrowski & G.M.Lieberman, *On the Dirichlet problem with L^2-boundary values in a half-space*, Indiana Univ. Math. J. **35** (1986), 623–642.

[CM] M.Chicco, *Solvability of the Dirichlet problem in $H^{2,p}(\Omega)$ for a class of linear second order elliptic partial differential equations*, Boll. Un. Mat. Ital. **4** (1971), 374–384.

[DA] B.E.J.Dahlberg, *On the absolute continuity of elliptic measures*, Am. J. of Maths **108** (1986), 1119–1138.

[**FE**] R.Fefferman, *A criterion for the absolute continuity of the harmonic measure associated with an elliptic operator*, J. of Am. Math. Soc. **2(1)** (1989), 127–135.

[**FR**] O.Frostman, *Potentiel d'équilibre et capacité des ensembles avec quelques applications à la théorie des fonctions*, Meddel. Lunds Univ. Mat. Sem. **3** (1935), 1–18.

[**FJK**] E.B.Fabes, D.S.Jerison & C.Kenig, *Necessary and sufficient conditions for absolute continuity of elliptic-harmonic measure*, Ann. of Maths **119** (1984), 121–141.

[**GA**] S.V.Gaǐdenko, *On exceptional sets on the boundary and the uniqueness of solutions of the Dirichlet problem for a second order elliptic equation*, Math. USSR Sbornik **39** (1981), 107–123.

[**GE**] F.W.Gehring, *The L^p-integrability of the partial derivatives of quasiconformal mapping*, Acta Math. **130** (1970), 265–277.

[**GI**] M.Giaquinta, *Multiple integrals in the calculus of variations and nonlinear elliptic systems*, Annals of Math. Studies, no.105, Princeton Univ. Press, Princeton, N.J., 1983.

[**GH**] D.Gilbarg & L.Hörmander, *Intermediate Schauder estimates*, Archiv. Rat. Mech. Anal. **74** (1980), 297–318.

[**GT**] D.Gilbarg & N.S.Trudinger, *Elliptic partial differential equations of second order*, Springer–Verlag, Berlin–Heidelberg–New York, 1983.

[**GM**] A.K.Guščin & V.P.Mihaǐlov, *On boundary values in L_p, $p > 1$, of solutions of elliptic equations*, Math. USSR Sbornik **36** (1980), 1–19.

[**GU**] A.K.Guščin, *On the Dirichlet problem for elliptic equation of the second order*, Mat. Sb. **137(179),No.9** (1988), 19–64.

[**HO**] L.Hörmander, *L^p-estimates for (pluri-) subharmonic functions*, Math. Scand. **20** (1967), 65–78.

[**HW1**] T.Hoffmann–Walbeck, *On the Dirichlet problem for linear elliptic equations and L^p-boundary data, $p > 1$*, Boll. U. Math. Ital. **7 I–B** (1987), 1–30.

[**HW2**] T.Hoffmann–Walbeck, *On the Dirichlet problem in the half-space with L^p -boundary data*, Boll. U. Math. Ital. **7 I**–B (1987), 889–904.

[**IT**] T.Iwaniec, *Gehring's reverse maximal function inequality*, Approximation and Function Spaces, Proceedings of the International Conference held in Gdańsk, 1979 (1981), 294–305, North–Holland Publishing Company, PWN–Polish Scientific Publishers, Amsterdam–N.Y.–Oxford, Warszawa.

[**JK**] D.Jerison & C.Kenig, *The Dirichlet problem in non-smooth domains*, Ann. of Maths **113** (1981), 367–382.

[**KO**] A.I.Koševelev, *Apriori L_p-estimates and generalized solutions of elliptic equations and systems*, Uspehi Mat. Nauk. **80** (1956), 29–88.

[**KR**] Steven G.Krantz, *Function theory of several complex variables*, A Wiley–Interscience Publication, New York, 1982.

[**KJF**] A.Kufner, O.John & S.Fučik, *Function spaces*, Noordhoof, Academia, Leyden –Prague, 1977.

[**KU**] A.Kufner, *Weighted Sobolev spaces*, Teubner– Texte zur Mathematik – Band 31, 1980.

[**KS**] A.Kufner & A.M.Sändig, *Some applications of weighted Sobolev spaces*, Teubner–Texte zur Mathematik – Band 100, 1987.

[**LA**] M.Langlais, *On the continuous solutions of a degenerate elliptic equation*, Proc. London Math. Soc. **3(50)** (1985), 282–298.

[LM] J.L.Lions & E.Magenes, *Problèmes aux limites non homogènes et applications, Vol. I*, Dunod, Paris, 1968.

[LU] O.A.Ladyzhenskaya & N.N.Ural'tseva, *Linear and quasilinear elliptic equations*, Izdat. Nauka, Moscow, 1968.

[LI1] G.M.Lieberman, *Regularized distance and its applications*, Pacific J. Math. **117** (1985), 329–352.

[LI2] G.M.Lieberman, *The Dirichlet problem with L^p boundary data in domains with Dini continuous normal*, to appear in Tsukuba J. Math.

[LP] J.E.Littlewood & R.E.A.C.Paley, *Theorems on Fourier series and power series (II)*, Proc. London Math. Soc. **42(2)** (1937), 52–89.

[LSW] N.Littman, G.Stampacchia & H.F.Weinberger, *Regular points for elliptic equations with discontinuous coefficients*, Ann. Scuola Norm. Sup. Pisa. **17** (1963), 47–79.

[MA] W.G.Maz'ja, Sobolev spaces, Springer-Verlag, Berlin-Heidelberg-Tokyo -New York.

[ME] R.D.Meyer, *Some embedding theorems for generalized Sobolev spaces and applications to degenerate elliptic differential operators*, J. Math. Mech. **16** (1967), 739–760.

[MM] L.Modica & S.Mortola, *Construction of a singular elliptic–harmonic measure*, Manuscripta Math. **33** (1980), 81–98.

[M1] V.P.Mihaĭlov, *The Dirichlet problem for a second order elliptic equation*, Differencial'nye Uravnenija **12** (1976), 1877–1891.

[M2] V.P.Mihaĭlov, *The boundary values of the solutions of the second order elliptic equations*, Mat. Sb. **100(142)** (1976), 5–13.

[M3] V.P.Mihaĭlov, *Boundary values of elliptic equations in domains with a smooth boundary*, Mat. Sb. **101(143)** (1976), 163–188.

[M4] V.P.Mihaĭlov, *On the boundary values of the solutions of elliptic equations*, Appl. Math. Optim. **6** (1980), 193–199.

[M5] V.P.Mihaĭlov, *Partial differential equations*, Izdat. Nauka, Moscow, 1976.

[MH] Yu.A.Mihaĭlov, *Solvability of the Dirichlet problem in the space H^1_{loc} for second–order elliptic equations (Russian)*, Differential Equations and Theory, Matematicheskie Issledovaniia (Akademia Nauk Moldavskoi SSR, Institut Matematiki **67** (1982), 109–123.

[MI] C.Miranda, *Partial differential equations of elliptic type*, Springer Verlag, New York, 1970.

[N1] J.Nečas, *On the regularity of second order elliptic partial differential equations with unbounded Dirichlet integral*, Arch. Rat. Mech. Anal. **9** (1962), 134–144.

[N2] J.Nečas, *Les méthodes directes en théorie des équations elliptiques*, Academia, Prague, 1967.

[OR] O.A.Oleĭnik & E.V.Radkevič, *Second order equations with non–negative characteristic form*, Am. Math. Soc. Providence, Plenum, New York, 1976.

[PE] I.M.Petrushko, *On the boundary values of solutions of elliptic equations in domains with Lyapunov boundary*, Math. USSR Sb. **47 (No.1)** (1984), 43–72.

[PE1] I.M.Petrushko, *On boundary values in L_p, $p > 1$, of solutions of elliptic equations in domains with a Lyapunov boundary*, Math. USSR Sb. **48 (No.2)** (1984), 565–585.

[FR] F.Riesz, *Über die Randwerte einer analytischen Funktion*, Math. Z. **18** (1923), 87–95.

[**ST1**] G.Stampacchia, *Le problème de Dirichlet pour les équations elliptiques du second ordre à coefficients discontinus*, Ann. Inst. Fourier (Grenoble) **15** (1965), 189–258.

[**ŠE**] W.Ju. Šelepow, *Properties of boundary values of solutions of elliptic equations in domains with boundaries representable in the form of the difference of two convex functions*, Mat. Sb. **133(175)** (1987), 446–468.

[**ST2**] G.Stampacchia, *Équations elliptiques du second ordre à coefficients discontinus*, Séminaire de Mathématiques Supérieures, 16, Les Presses de l'Université de Montrál, Montréal, Quebec, 1965.

[**ST**] E.M.Stein, *Singular integrals and differentiability properties of functions*, Princeton University Press, Princeton, New Yersey, 1970.

[**TA**] G.Talenti, *Best constant in Sobolev inequality*, Ann. Mat. Pura Appl. **110** (1976), 352–372.

[**TT**] Maria Transirico & Mario Troisi, *Equazioni ellittiche del secondo ordine a coefficienti discontinui e di tipo variazionale in aperti non limitati*, Boll. U. M. I. **(7) 2-B** (1988), 385–398.

[**TR**] Mario Troisi, *Su una classe di spazi di Sobolev con peso*, Rend. Accad. Naz. Sci. XL Mem. Mat. Sci. Fis. Natur. **(15)10(104)** (1986), 177–189.

INDEX

Lecture Notes in Mathematics

For information about Vols. 1–1296
please contact your bookseller or Springer-Verlag

Vol. 1297: Zhu Y.-I., Guo B.-y. (Eds.), Numerical Methods for Partial Differential Equations. Proceedings. XI, 244 pages. 1987.

Vol. 1298: J. Aguadé, R. Kane (Eds.), Algebraic Topology, Barcelona 1986. Proceedings. X, 255 pages. 1987.

Vol. 1299: S. Watanabe, Yu.V. Prokhorov (Eds.), Probability Theory and Mathematical Statistics. Proceedings, 1986. VIII, 589 pages. 1988.

Vol. 1300: G.B. Seligman, Constructions of Lie Algebras and their Modules. VI, 190 pages. 1988.

Vol. 1301: N. Schappacher, Periods of Hecke Characters. XV, 160 pages. 1988.

Vol. 1302: M. Cwikel, J. Peetre, Y. Sagher, H. Wallin (Eds.), Function Spaces and Applications. Proceedings, 1986. VI, 445 pages. 1988.

Vol. 1303: L. Accardi, W. von Waldenfels (Eds.), Quantum Probability and Applications III. Proceedings, 1987. VI, 373 pages. 1988.

Vol. 1304: F.Q. Gouvêa, Arithmetic of p-adic Modular Forms. VIII, 121 pages. 1988.

Vol. 1305: D.S. Lubinsky, E.B. Saff, Strong Asymptotics for Extremal Polynomials Associated with Weights on R. VII, 153 pages. 1988.

Vol. 1306: S.S. Chern (Ed.), Partial Differential Equations. Proceedings, 1986. VI, 294 pages. 1988.

Vol. 1307: T. Murai, A Real Variable Method for the Cauchy Transform, and Analytic Capacity. VIII, 133 pages. 1988.

Vol. 1308: P. Imkeller, Two-Parameter Martingales and Their Quadratic Variation. IV, 177 pages. 1988.

Vol. 1309: B. Fiedler, Global Bifurcation of Periodic Solutions with Symmetry. VIII, 144 pages. 1988.

Vol. 1310: O.A. Laudal, G. Pfister: Local Moduli and Singularities. V, 117 pages. 1988.

Vol. 1311: A. Holme, R. Speiser (Eds.), Algebraic Geometry, Sundance 1986. Proceedings. VI, 320 pages. 1988.

Vol. 1312: N.A. Shirokov, Analytic Functions Smooth up to the Boundary. III, 213 pages. 1988.

Vol. 1313: F. Colonius. Optimal Periodic Control. VI, 177 pages. 1988.

Vol. 1314: A. Futaki. Kähler-Einstein Metrics and Integral Invariants. IV, 140 pages. 1988.

Vol. 1315: R.A. McCoy, I. Ntantu, Topological Properties of Spaces of Continuous, Functions. IV, 124 pages. 1988.

Vol. 1316: H. Korezlioglu, A.S. Ustunel (Eds.), Stochastic Analysis and Related Topics. Proceedings, 1986. V, 371 pages. 1988.

Vol. 1317: J. Lindenstrauss, V.D. Milman (Eds.), Geometric Aspects of Functional Analysis. Seminar, 1986–87. VII, 289 pages. 1988.

Vol. 1318: Y. Felix (Ed.), Algebraic Topology – Rational Homotopy. Proceedings, 1986. VIII, 245 pages. 1988.

Vol. 1319: M. Vuorinen, Conformal Geometry and Quasiregular Mappings. XIX, 209 pages. 1988.

Vol. 1320: H. Jürgensen, G. Lallement, H.J. Weinert (Eds.), Semigroups, Theory and Applications. Proceedings, 1986. X, 416 pages. 1988.

Vol. 1321: J. Azéma, P.A. Meyer, M. Yor (Eds.), Séminaire de Probabilités XXII. Proceedings. IV, 600 pages. 1988.

Vol. 1322: M. Métivier, S. Watanabe (Eds.), Stochastic Analysis. Proceedings, 1987. VII, 197 pages. 1988.

Vol. 1323: D.R. Anderson, H.J. Munkholm, Boundedly Controlled Topology. XII, 309 pages. 1988.

Vol. 1324: F. Cardoso, D.G. de Figueiredo, R. Iório, O. Lopes (Eds.), Partial Differential Equations. Proceedings, 1986. VIII, 433 pages. 1988.

Vol. 1325: A. Truman, I.M. Davies (Eds.), Stochastic Mechanics and Stochastic Processes. Proceedings, 1986. V, 220 pages. 1988.

Vol. 1326: P.S. Landweber (Ed.), Elliptic Curves and Modular Forms in Algebraic Topology. Proceedings, 1986. V, 224 pages. 1988.

Vol. 1327: W. Bruns, U. Vetter, Determinantal Rings. VII, 236 pages. 1988.

Vol. 1328: J.L. Bueso, P. Jara, B. Torrecillas (Eds.), Ring Theory. Proceedings, 1986. IX, 331 pages. 1988.

Vol. 1329: M. Alfaro, J.S. Dehesa, F.J. Marcellan, J.L. Rubio de Francia, J. Vinuesa (Eds.): Orthogonal Polynomials and their Applications. Proceedings, 1986. XV, 334 pages. 1988.

Vol. 1330: A. Ambrosetti, F. Gori, R. Lucchetti (Eds.), Mathematical Economics. Montecatini Terme 1986. Seminar. VII, 137 pages. 1988.

Vol. 1331: R. Bamón, R. Labarca, J. Palis Jr. (Eds.), Dynamical Systems, Valparaiso 1986. Proceedings. VI, 250 pages. 1988.

Vol. 1332: E. Odell, H. Rosenthal (Eds.), Functional Analysis. Proceedings. 1986–87. V, 202 pages. 1988.

Vol. 1333: A.S. Kechris, D.A. Martin, J.R. Steel (Eds.), Cabal Seminar 81–85. Proceedings, 1981–85. V, 224 pages. 1988.

Vol. 1334: Yu.G. Borisovich, Yu.E. Gliklikh (Eds.), Global Analysis – Studies and Applications III. V, 331 pages. 1988.

Vol. 1335: F. Guillén, V. Navarro Aznar, P. Pascual-Gainza, F. Puerta, Hyperrésolutions cubiques et descente cohomologique. XII, 192 pages. 1988.

Vol. 1336: B. Helffer, Semi-Classical Analysis for the Schrödinger Operator and Applications. V, 107 pages. 1988.

Vol. 1337: E. Sernesi (Ed.), Theory of Moduli. Seminar, 1985. VIII, 232 pages. 1988.

Vol. 1338: A.B. Mingarelli, S.G. Halvorsen. Non-Oscillation Domains of Differential Equations with Two Parameters. XI, 109 pages. 1988.

Vol. 1339: T. Sunada (Ed.), Geometry and Analysis of Manifolds. Proceedings, 1987. IX, 277 pages. 1988.

Vol. 1340: S. Hildebrandt, D.S. Kinderlehrer, M. Miranda (Eds.), Calculus of Variations and Partial Differential Equations. Proceedings, 1986. IX, 301 pages. 1988.

Vol. 1341: M. Dauge, Elliptic Boundary Value Problems on Corner Domains. VIII, 259 pages. 1988.

Vol. 1342: J.C. Alexander (Ed.), Dynamical Systems. Proceedings, 1986–87. VIII, 726 pages. 1988.

Vol. 1343: H. Ulrich, Fixed Point Theory of Parametrized Equivariant Maps. VII, 147 pages. 1988.

Vol. 1344: J. Král, J. Lukes, J. Netuka, J. Vesely´ (Eds.), Potential Theory – Surveys and Problems. Proceedings, 1987. VIII, 271 pages. 1988.

Vol. 1345: X. Gomez-Mont, J. Seade, A. Verjovski (Eds.), Holomorphic Dynamics. Proceedings, 1986. VII. 321 pages. 1988.

Vol. 1346: O.Ya. Viro (Ed.), Topology and Geometry – Rohlin Seminar. XI, 581 pages. 1988.

Vol. 1347: C. Preston, Iterates of Piecewise Monotone Mappings on an Interval. V, 166 pages. 1988.

Vol. 1348: F. Borceux (Ed.), Categorical Algebra and its Applications. Proceedings, 1987. VIII, 375 pages. 1988.

Vol. 1349: E. Novak, Deterministic and Stochastic Error Bounds in Numerical Analysis. V, 113 pages. 1988.

Vol. 1350: U. Koschorke (Ed.), Differential Topology Proceedings, 1987, VI, 269 pages. 1988.

Vol. 1351: I. Laine, S. Rickman, T. Sorvali (Eds.), Complex Analysis, Joensuu 1987. Proceedings. XV, 378 pages. 1988.

Vol. 1352: L.L. Avramov, K.B. Tchakerian (Eds.), Algebra – Some Current Trends. Proceedings. 1986. IX, 240 Seiten. 1988.

Vol. 1353: R.S. Palais, Ch.-l. Teng, Critical Point Theory and Submanifold Geometry. X, 272 pages. 1988.

Vol. 1354: A. Gómez, F. Guerra, M.A. Jiménez, G. López (Eds.), Approximation and Optimization. Proceedings, 1987. VI, 280 pages. 1988.

Vol. 1355: J. Bokowski, B. Sturmfels, Computational Synthetic Geometry. V, 168 pages. 1989.

Vol. 1356: H. Volkmer, Multiparameter Eigenvalue Problems and Expansion Theorems. VI, 157 pages. 1988.

Vol. 1357: S. Hildebrandt, R. Leis (Eds.), Partial Differential Equations and Calculus of Variations. VI, 423 pages. 1988.

Vol. 1358: D. Mumford, The Red Book of Varieties and Schemes. V, 309 pages. 1988.

Vol. 1359: P. Eymard, J.-P. Pier (Eds.) Harmonic Analysis. Proceedings, 1987. VIII, 287 pages. 1988.

Vol. 1360: G. Anderson, C. Greengard (Eds.), Vortex Methods. Proceedings, 1987. V, 141 pages. 1988.

Vol. 1361: T. tom Dieck (Ed.), Algebraic Topology and Transformation Groups. Proceedings. 1987. VI, 298 pages. 1988.

Vol. 1362: P. Diaconis, D. Elworthy, H. Föllmer, E. Nelson, G.C. Papanicolaou, S.R.S. Varadhan. École d´ Été de Probabilités de Saint-Flour XV–XVII. 1985–87 Editor: P.L. Hennequin. V, 459 pages. 1988.

Vol. 1363: P.G. Casazza, T.J. Shura, Tsirelson´s Space. VIII, 204 pages. 1988.

Vol. 1364: R.R. Phelps, Convex Functions, Monotone Operators and Differentiability. IX, 115 pages. 1989.

Vol. 1365: M. Giaquinta (Ed.), Topics in Calculus of Variations. Seminar, 1987. X, 196 pages. 1989.

Vol. 1366: N. Levitt, Grassmannians and Gauss Maps in PL-Topology. V, 203 pages. 1989.

Vol. 1367: M. Knebusch, Weakly Semialgebraic Spaces. XX, 376 pages. 1989.

Vol. 1368: R. Hübl, Traces of Differential Forms and Hochschild Homology. III, 111 pages. 1989.

Vol. 1369: B. Jiang, Ch.-K. Peng, Z. Hou (Eds.), Differential Geometry and Topology. Proceedings, 1986–87. VI, 366 pages. 1989.

Vol. 1370: G. Carlsson, R.L. Cohen, H.R. Miller, D.C. Ravenel (Eds.), Algebraic Topology. Proceedings, 1986. IX, 456 pages. 1989.

Vol. 1371: S. Glaz, Commutative Coherent Rings. XI, 347 pages. 1989.

Vol. 1372: J. Azéma, P.A. Meyer, M. Yor (Eds.), Séminaire de Probabilités XXIII. Proceedings. IV, 583 pages. 1989.

Vol. 1373: G. Benkart, J.M. Osborn (Eds.), Lie Algebras. Madison 1987. Proceedings. V, 145 pages. 1989.

Vol. 1374: R.C. Kirby, The Topology of 4-Manifolds. VI, 108 pages. 1989.

Vol. 1375: K. Kawakubo (Ed.), Transformation Groups. Proceedings, 1987. VIII, 394 pages, 1989.

Vol. 1376: J. Lindenstrauss, V.D. Milman (Eds.), Geometric Aspects of Functional Analysis. Seminar (GAFA) 1987–88. VII, 288 pages. 1989.

Vol. 1377: J.F. Pierce, Singularity Theory, Rod Theory, and Symmetry-Breaking Loads. IV, 177 pages. 1989.

Vol. 1378: R.S. Rumely, Capacity Theory on Algebraic Curves. III, 437 pages. 1989.

Vol. 1379: H. Heyer (Ed.), Probability Measures on Groups IX. Proceedings, 1988. VIII, 437 pages. 1989.

Vol. 1380: H.P. Schlickewei, E. Wirsing (Eds.), Number Theory, Ulm 1987. Proceedings. V, 266 pages. 1989.

Vol. 1381: J.-O. Strömberg, A. Torchinsky, Weighted Hardy Spaces. V, 193 pages. 1989.

Vol. 1382: H. Reiter, Metaplectic Groups and Segal Algebras. XI, 128 pages. 1989.

Vol. 1383: D.V. Chudnovsky, G.V. Chudnovsky, H. Cohn, M.B. Nathanson (Eds.), Number Theory, New York 1985–88. Seminar. V, 256 pages. 1989.

Vol. 1384: J. Garcia-Cuerva (Ed.), Harmonic Analysis and Partial Differential Equations. Proceedings, 1987. VII, 213 pages. 1989.

Vol. 1385: A.M. Anile, Y. Choquet-Bruhat (Eds.), Relativistic Fluid Dynamics. Seminar, 1987. V, 308 pages. 1989.

Vol. 1386: A. Bellen, C.W. Gear, E. Russo (Eds.), Numerical Methods for Ordinary Differential Equations. Proceedings, 1987. VII, 136 pages. 1989.

Vol. 1387: M. Petkovi´c, Iterative Methods for Simultaneous Inclusion of Polynomial Zeros. X, 263 pages. 1989.

Vol. 1388: J. Shinoda, T.A. Slaman, T. Tugué (Eds.), Mathematical Logic and Applications. Proceedings, 1987. V, 223 pages. 1989.

Vol. 1000: Second Edition. H. Hopf, Differential Geometry in the Large. VII, 184 pages. 1989.

Vol. 1389: E. Ballico, C. Ciliberto (Eds.), Algebraic Curves and Projective Geometry. Proceedings, 1988. V, 288 pages. 1989.

Vol. 1390: G. Da Prato, L. Tubaro (Eds.), Stochastic Partial Differential Equations and Applications II. Proceedings, 1988. VI, 258 pages. 1989.

Vol. 1391: S. Cambanis, A. Weron (Eds.), Probability Theory on Vector Spaces IV. Proceedings, 1987. VIII, 424 pages. 1989.

Vol. 1392: R. Silhol, Real Algebraic Surfaces. X, 215 pages. 1989.

Vol. 1393: N. Bouleau, D. Feyel, F. Hirsch, G. Mokobodzki (Eds.), Séminaire de Théorie du Potentiel Paris, No. 9. Proceedings. VI, 265 pages. 1989.

Vol. 1394: T.L. Gill, W.W. Zachary (Eds.), Nonlinear Semigroups, Partial Differential Equations and Attractors. Proceedings, 1987. IX, 233 pages. 1989.

Vol. 1395: K. Alladi (Ed.), Number Theory, Madras 1987. Proceedings. VII, 234 pages. 1989.

Vol. 1396: L. Accardi, W. von Waldenfels (Eds.), Quantum Probability and Applications IV. Proceedings, 1987. VI, 355 pages. 1989.

Vol. 1397: P.R. Turner (Ed.), Numerical Analysis and Parallel Processing. Seminar, 1987. VI, 264 pages. 1989.

Vol. 1398: A.C. Kim, B.H. Neumann (Eds.), Groups – Korea 1988. Proceedings. V, 189 pages. 1989.

Vol. 1399: W.-P. Barth, H. Lange (Eds.), Arithmetic of Complex Manifolds. Proceedings, 1988. V, 171 pages. 1989.

Vol. 1400: U. Jannsen. Mixed Motives and Algebraic K-Theory. XIII, 246 pages. 1990.

Vol. 1401: J. Steprans, S. Watson (Eds.), Set Theory and its Applications. Proceedings, 1987. V, 227 pages. 1989.

Vol. 1402: C. Carasso, P. Charrier, B. Hanouzet, J.-L. Joly (Eds.), Nonlinear Hyperbolic Problems. Proceedings, 1988. V, ? pages. 1989.

Vol. 1403: B. Simeone (Ed.), Combinatorial Optimization. Seminar, 1986. V, 314 pages. 1989.

Vol. 1404: M.-P. Malliavin (Ed.), Séminaire d'Algèbre Paul Dubreil et Marie-Paul Malliavin. Proceedings, 1987–1988. IV, pages. 1989.

Vol. 1405: S. Dolecki (Ed.), Optimization. Proceedings, 1988. ?23 pages. 1989. Vol. 1406: L. Jacobsen (Ed.), Analytic Theory of Continued Fractions III. Proceedings, 1988. VI, 142 pages. 1989.

Vol. 1407: W. Pohlers, Proof Theory. VI, 213 pages. 1989.

Vol. 1408: W. Lück, Transformation Groups and Algebraic K-Theory. XII, 443 pages. 1989.

Vol. 1409: E. Hairer, Ch. Lubich, M. Roche. The Numerical Solution of Differential-Algebraic Systems by Runge-Kutta Methods. VII, 139 pages. 1989.

Vol. 1410: F.J. Carreras, O. Gil-Medrano, A.M. Naveira (Eds.), Differential Geometry. Proceedings, 1988. V, 308 pages. 1989.

Vol. 1411: B. Jiang (Ed.), Topological Fixed Point Theory and Applications. Proceedings. 1988. VI, 203 pages. 1989.

Vol. 1412: V.V. Kalashnikov, V.M. Zolotarev (Eds.), Stability Problems for Stochastic Models. Proceedings, 1987. X, 380 pages. 1989.

Vol. 1413: S. Wright, Uniqueness of the Injective III_1 Factor. III, pages. 1989.

Vol. 1414: E. Ramirez de Arellano (Ed.), Algebraic Geometry and Complex Analysis. Proceedings, 1987. VI, 180 pages. 1989.

Vol. 1415: M. Langevin, M. Waldschmidt (Eds.), Cinquante Ans de Polynômes. Fifty Years of Polynomials. Proceedings, IX, 235 pages. 1990.

Vol. 1416: C. Albert (Ed.), Géométrie Symplectique et ...ique. Proceedings, 1988. V, 289 pages. 1990.

Vol. 1417: A.J. Sommese, A. Biancofiore, E.L. Livorni (Eds.), ...ic Geometry. Proceedings, 1988. V, 320 pages. 1990.

Vol. 1418: M. Mimura (Ed.), Homotopy Theory and Related ... Proceedings, 1988. V, 241 pages. 1990.

Vol. 1419: P.S. Bullen, P.Y. Lee, J.L. Mawhin, P. Muldowney, ...r (Eds.), New Integrals. Proceedings, 1988. V, 202 ...

Vol. 1420: M. Galbiati, A. Tognoli (Eds.), Real Analytic ... Proceedings, 1988. IV, 366 pages. 1990.

Vol. 1421: H.A. Biagioni, A Nonlinear Theory of Generalized Functions. XII, 214 pages. 1990.

Vol. 1422: V. Villani (Ed.), Complex Geometry and Analysis. Proceedings, 1988. V, 109 pages. 1990.

Vol. 1423: S.O. Kochman, Stable Homotopy Groups of Spheres: A Computer-Assisted Approach. VIII, 330 pages. 1990.

Vol. 1424: F.E. Burstall, J.H. Rawnsley, Twistor Theory for Riemannian Symmetric Spaces. III, 112 pages. 1990.

Vol. 1425: R.A. Piccinini (Ed.), Groups of Self-Equivalences and Related Topics. Proceedings, 1988. V, 214 pages. 1990.

Vol. 1426: J. Azéma, P.A. Meyer, M. Yor (Eds.), Séminaire de Probabilités XXIV, 1988/89. V, 490 pages. 1990.

Vol. 1427: A. Ancona, D. Geman, N. Ikeda, École d'Eté de Probabilités de Saint Flour XVIII, 1988. Ed.: P.L. Hennequin. VII, 330 pages. 1990.

Vol. 1428: K. Erdmann, Blocks of Tame Representation Type and Related Algebras. XV. 312 pages. 1990.

Vol. 1429: S. Homer, A. Nerode, R.A. Platek, G.E. Sacks, A. Scedrov, Logic and Computer Science. Seminar, 1988. Editor: P. Odifreddi. V, 162 pages. 1990.

Vol. 1430: W. Bruns, A. Simis (Eds.), Commutative Algebra. Proceedings. 1988. V, 160 pages. 1990.

Vol. 1431: J.G. Heywood, K. Masuda, R. Rautmann, V.A. Solonnikov (Eds.), The Navier-Stokes Equations – Theory and Numerical Methods. Proceedings, 1988. VII, 238 pages. 1990.

Vol. 1432: K. Ambos-Spies, G.H. Müller, G.E. Sacks (Eds.), Recursion Theory Week. Proceedings, 1989. VI, 393 pages. 1990.

Vol. 1433: S. Lang, W. Cherry, Topics in Nevanlinna Theory. II, 174 pages.1990.

Vol. 1434: K. Nagasaka, E. Fouvry (Eds.), Analytic Number Theory. Proceedings, 1988. VI, 218 pages. 1990.

Vol. 1435: St. Ruscheweyh, E.B. Saff, L.C. Salinas, R.S. Varga (Eds.), Computational Methods and Function Theory. Proceedings, 1989. VI, 211 pages. 1990.

Vol. 1436: S. Xambó-Descamps (Ed.), Enumerative Geometry. Proceedings, 1987. V, 303 pages. 1990.

Vol. 1437: H. Inassaridze (Ed.), K-theory and Homological Algebra. Seminar, 1987–88. V, 313 pages. 1990.

Vol. 1438: P.G. Lemarié (Ed.) Les Ondelettes en 1989. Seminar. IV, 212 pages. 1990.

Vol. 1439: E. Bujalance, J.J. Etayo, J.M. Gamboa, G. Gromadzki. Automorphism Groups of Compact Bordered Klein Surfaces: A Combinatorial Approach. XIII, 201 pages. 1990.

Vol. 1440: P. Latiolais (Ed.), Topology and Combinatorial Groups Theory. Seminar, 1985–1988. VI, 207 pages. 1990.

Vol. 1441: M. Coornaert, T. Delzant, A. Papadopoulos. Géométrie et théorie des groupes. X, 165 pages. 1990.

Vol. 1442: L. Accardi, M. von Waldenfels (Eds.), Quantum Probability and Applications V. Proceedings, 1988. VI, 413 pages. 1990.

Vol. 1443: K.H. Dovermann, R. Schultz, Equivariant Surgery Theories and Their Periodicity Properties. VI, 227 pages. 1990.

Vol. 1444: H. Korezlioglu, A.S. Ustunel (Eds.), Stochastic Analysis and Related Topics VI. Proceedings, 1988. V, 268 pages. 1990.

Vol. 1445: F. Schulz, Regularity Theory for Quasilinear Elliptic Systems and – Monge Ampère Equations in Two Dimensions. XV, 123 pages. 1990.

Vol. 1446: Methods of Nonconvex Analysis. Seminar, 1989. Editor: A. Cellina. V, 206 pages. 1990.

Vol. 1447: J.-G. Labesse, J. Schwermer (Eds), Cohomology of Arithmetic Groups and Automorphic Forms. Proceedings, 1989. V, 358 pages. 1990.

Vol. 1448: S.K. Jain, S.R. López-Permouth (Eds.), Non-Commutative Ring Theory. Proceedings, 1989. V, 166 pages. 1990.

Vol. 1449: W. Odyniec, G. Lewicki, Minimal Projections in Banach Spaces. VIII, 168 pages. 1990.

Vol. 1450: H. Fujita, T. Ikebe, S.T. Kuroda (Eds.), Functional-Analytic Methods for Partial Differential Equations. Proceedings, 1989. VII, 252 pages. 1990.

Vol. 1451: L. Alvarez-Gaumé, E. Arbarello, C. De Concini, N.J. Hitchin, Global Geometry and Mathematical Physics. Montecatini Terme 1988. Seminar. Editors: M. Francaviglia, F. Gherardelli. IX, 197 pages. 1990.

Vol. 1452: E. Hlawka, R.F. Tichy (Eds.), Number-Theoretic Analysis. Seminar, 1988–89. V, 220 pages. 1990.

Vol. 1453: Yu.G. Borisovich, Yu.E. Gliklikh (Eds.), Global Analysis – Studies and Applications IV. V, 320 pages. 1990.

Vol. 1454: F. Baldassari, S. Bosch, B. Dwork (Eds.), p-adic Analysis. Proceedings, 1989. V, 382 pages. 1990.

Vol. 1455: J.-P. Françoise, R. Roussarie (Eds.), Bifurcations of Planar Vector Fields. Proceedings, 1989. VI, 396 pages. 1990.

Vol. 1456: L.G. Kovács (Ed.), Groups – Canberra 1989. Proceedings. XII, 198 pages. 1990.

Vol. 1457: O. Axelsson, L.Yu. Kolotilina (Eds.), Preconditioned Conjugate Gradient Methods. Proceedings, 1989. V, 196 pages. 1990.

Vol. 1458: R. Schaaf, Global Solution Branches of Two Point Boundary Value Problems. XIX, 141 pages. 1990.

Vol. 1459: D. Tiba, Optimal Control of Nonsmooth Distributed Parameter Systems. VII, 159 pages. 1990.

Vol. 1460: G. Toscani, V. Boffi, S. Rionero (Eds.), Mathematical Aspects of Fluid Plasma Dynamics. Proceedings, 1988. V, 221 pages. 1991.

Vol. 1461: R. Gorenflo, S. Vessella, Abel Integral Equations. VII, 215 pages. 1991.

Vol. 1462: D. Mond, J. Montaldi (Eds.), Singularity Theory and its Applications. Warwick 1989, Part I. VIII, 405 pages. 1991.

Vol. 1463: R. Roberts, I. Stewart (Eds.), Singularity Theory and its Applications. Warwick 1989, Part II. VIII, 322 pages. 1991.

Vol. 1464: D. L. Burkholder, E. Pardoux, A. Sznitman, Ecole d'Eté de Probabilités de Saint- Flour XIX-1989. Editor: P. L. Hennequin. VI, 256 pages. 1991.

Vol. 1465: G. David, Wavelets and Singular Integrals on Curves and Surfaces. X, 107 pages. 1991.

Vol. 1466: W. Banaszczyk, Additive Subgroups of Topological Vector Spaces. VII, 178 pages. 1991.

Vol. 1467: W. M. Schmidt, Diophantine Approximations and Diophantine Equations. VIII, 217 pages. 1991.

Vol. 1468: J. Noguchi, T. Ohsawa (Eds.), Prospects in Complex Geometry. Proceedings, 1989. VII, 421 pages. 1991.

Vol. 1469: J. Lindenstrauss, V. D. Milman (Eds.), Geometric Aspects of Functional Analysis. Seminar 1989-90. XI, 191 pages. 1991.

Vol. 1470: E. Odell, H. Rosenthal (Eds.), Functional Analysis. Proceedings, 1987-89. VII, 199 pages. 1991.

Vol. 1471: A. A. Panchishkin, Non-Archimedean L-Functions of Siegel and Hilbert Modular Forms. VII, 157 pages. 1991.

Vol. 1472: T. T. Nielsen, Bose Algebras: The Complex and Real Wave Representations. V, 132 pages. 1991.

Vol. 1473: Y. Hino, S. Murakami, T. Naito, Functional Diff- rential Equations with Infinite Delay. X, 317 pages. 1991.

Vol. 1474: S. Jackowski, B. Oliver, K. Pawałowski (Eds. Algebraic Topology, Poznań 1989. Proceedings. VIII, 397 page 1991.

Vol. 1475: S. Busenberg, M. Martelli (Eds.), Delay Differenti Equations and Dynamical Systems. Proceedings, 1990. VIII, 24 pages. 1991.

Vol. 1476: M. Bekkali, Topics in Set Theory. VII, 120 page 1991.

Vol. 1477: R. Jajte, Strong Limit Theorems in Noncommutati L_2-Spaces. X, 113 pages. 1991.

Vol. 1478: M.-P. Malliavin (Ed.), Topics in Invariant Theo Seminar 1989-1990. VI, 272 pages. 1991.

Vol. 1479: S. Bloch, I. Dolgachev, W. Fulton (Eds.), Algebr Geometry. Proceedings, 1989. VII, 300 pages. 1991.

Vol. 1482: J. Chabrowski, The Dirichlet Problem with Boundary Data for Elliptic Linear Equations. VI, 173 pa 1991.